Advances in Experimental Medicine and Biology

Volume 1181

More information about this series at http://www.springer.com/series/5584

Zhibin Lin • Baoxue Yang
Editors

Ganoderma and Health

Biology, Chemistry and Industry

Springer

Editors
Zhibin Lin
Department of Pharmacology
School of Basic Medical Sciences,
Peking University
Beijing, China

Baoxue Yang
Department of Pharmacology
School of Basic Medical Sciences,
Peking University
Beijing, China

ISSN 0065-2598 ISSN 2214-8019 (electronic)
Advances in Experimental Medicine and Biology
ISBN 978-981-13-9869-8 ISBN 978-981-13-9867-4 (eBook)
https://doi.org/10.1007/978-981-13-9867-4

This Springer imprint is published by the registered company Springer Nature Singapore Pte Ltd.
The registered company address is: 152 Beach Road, #21-01/04 Gateway East, Singapore 189721, Singapore

Contents

Contributors

Hongyu Chen Institute of Edible Fungi, Shanghai Academy of Agricultural Sciences, Shanghai, China

Ruoyun Chen Department of Natural Product Chemistry, Institute of Materia Medica, Chinese Academy of Medical Sciences & Peking Union Medical College, Beijing, China

Cai-Hong Dong State Key Laboratory of Mycology, Institute of Microbiology, Chinese Academy of Sciences, Beijing, China

Zhuo Du State Key Laboratory of Mycology, Institute of Microbiology, Chinese Academy of Sciences, Beijing, China

Jie Feng Institute of Edible Fungi, Shanghai Academy of Agricultural Sciences, Shanghai, China

Ting Gong Department of Natural Product Chemistry, Institute of Materia Medica, Chinese Academy of Medical Sciences & Peking Union Medical College, Beijing, China

Jie Kang Department of Natural Product Chemistry, Institute of Materia Medica, Chinese Academy of Medical Sciences & Peking Union Medical College, Beijing, China

Peng Li Department of Pharmacology, School of Pharmacy, Fujian Provincial Key Laboratory of Natural Medicine Pharmacology, Fujian Medical University, Fuzhou, Fujian, China

Zhenhao Li Zhejiang Shouxiangu Pharmaceutical Co., Ltd, Wuyi, Zhejiang, China
College of Pharmaceutical Sciences, Zhejiang University, Hangzhou, Zhejiang, China

Zhibin Lin Department of Pharmacology, School of Basic Medical Sciences, Peking University, Beijing, China

Yanfang Liu Institute of Edible Fungi, Shanghai Academy of Agricultural Sciences, Shanghai, China

Qingjiu Tang Institute of Edible Fungi, Shanghai Academy of Agricultural Sciences, Shanghai, China

Ke Wang State Key Laboratory of Mycology, Institute of Microbiology, Chinese Academy of Sciences, Beijing, China

Jianhua Xu Department of Pharmacology, School of Pharmacy, Fujian Provincial Key Laboratory, of Natural Medicine Pharmacology, Fujian Medical University, Fuzhou, Fujian, China

Renyi Yan Department of Natural Product Chemistry, Institute of Materia Medica, Chinese Academy of Medical Sciences & Peking Union Medical College, Beijing, China

Yi-Jian Yao State Key Laboratory of Mycology, Institute of Microbiology, Chinese Academy of Sciences, Beijing, China

Jingsong Zhang Institute of Edible Fungi, Shanghai Academy of Agricultural Sciences, Shanghai, China

Jianlong Zhou Zhejiang Shouxiangu Pharmaceutical Co., Ltd, Wuyi, Zhejiang, China

Shuai Zhou Institute of Edible Fungi, Shanghai Academy of Agricultural Sciences, Shanghai, China

Chapter 1
Ganoderma (Lingzhi) in Traditional Chinese Medicine and Chinese Culture

Zhibin Lin

Abstract *Shen Nong Ben Cao Jing (Shennong Materia Medica)* and many other books in early Chinese history began to study, discuss, and report the scientific aspects of *Ganoderma* (Lingzhi) in respect to its categorization, habitat, bionomics, herbal nature, medication, etc. At the same time, incorrect or unsubstantiated information continues to be weeded out and updated. *Shennong Materia Medica* have been frequently referred in literature and used for further research and applications. Present chapter reviews the history of modern research on *Ganoderma* (Lingzhi) since 1950s.

Historically, Lingzhi has been viewed as a magic herb as well as an auspicious symbol by the Chinese. It is, therefore, also known as "Ruizhi," "Shenzhi," and "Xiancao," with the meaning of good fortune and mysterious power. Taoism played an important role in promoting Lingzhi for either medical purposes or otherwise. Numerous myths and poems mentioning people's love, worshipping, and beliefs on Lingzhi can be found in the Chinese literature since ancient times.

Keywords *Ganoderma* · Lingzhi · History · Culture · Myths

1.1 *Ganoderma* (Lingzhi) in Traditional Chinese Medicine

1.1.1 Ganoderma *(Lingzhi) Described in Shen Nong Ben Cao Jing*

Written in ca. 100 B.C., *Shen Nong Ben Cao Jing* (*Shennong Materia Medica*) was the earliest pharmaceutical book of Lingzhi. *Shennong Materia Medica* was the summation of the experiences gained by the ancient Chinese medical practitioners. The legend of the God of Agriculture, who tasted a hundred herbs and encountered

Z. Lin (✉)
Department of Pharmacology, School of Basic Medical Sciences,
Peking University, Beijing, China
e-mail: linzb1937@sina.com

Z. Lin, B. Yang (eds.), *Ganoderma and Health*, Advances in Experimental Medicine and Biology 1181, https://doi.org/10.1007/978-981-13-9867-4_1

Fig. 1.1 God of agriculture. A Liao Dynasty era painting on the wall of a wooden pagoda in Shanxi showing a barefooted man with a plump face and bared belly, covered himself with fur and leaves, shouldered a bamboo basket, held a Lingzhi, and walked on a rocky mountain. Scholars believe that the man in the picture was Shennong, the God of Agriculture. (Reproduced with permission from Ref. [2])

70 poisonous ones in a day, gives the discovery process a vivid picture. Likewise, the medicinal information on Lingzhi shown in *Shennong Materia Medica* was obtained through practices (Fig. 1.1). The book's real author is not known. Nonetheless, this monumental treatise listed 365 medicinal materials in categories according to their medicinal and toxic effects on human. The highest graded materials were those with medicinal efficacies without known toxicity. Lingzhi (including Cizhi, Qingzhi, Huangzhi, Baizhi, Heizhi, and Zizhi, which are red, blue, yellow, white, black, and purple, respectively) was among the highest graded materials, based on the categorization.

According to the theory of Yin and Yang and the 5-Elements in traditional Chinese medicine (TCM), *Shennong Materia Medica* categorized Lingzhi according to their colors into Cizhi (or Danzhi), Heizhi (or Xuanzhi), Qingzhi (or Longzhi), Baizhi (or Yuzhi), and Huangzhi (or Jinzhi) as "Five Zhi," as well as Zizhi (or Muzhi). Detailed descriptions on their medicinal nature, properties and flavors, and medication were included in the book, indicating that Cizhi was bitter in taste, mild in nature, and nontoxic and could be used for relieving chest congestion and improving memory; Heizhi was salty, mild, and nontoxic and could be used for renal problems and increasing awareness; Qingzhi was acidic, mild, and nontoxic and could be used to improve eyesight and liver functions; Baizhi was pungent, mild, and nontoxic and could be used to cure coughs and ailments of the lungs; Huangzhi was sweet, mild, and nontoxic and could be used to rid of heart, spleen, and stomach illnesses; and Zizhi was sweet, warm (mild), and nontoxic and could be used to treat hard-hearing and arthritis. In addition, it emphasized that all six kinds of Lingzhi could be applied for long term to facilitate health, well-being, and longevity. Such information on Lingzhi is considered the classic reference for the TCM and has been quoted frequently in the literatures to date [1].

1.1.2 *Expounding on* Ganoderma *(Lingzhi) by Ancient Chinese Scholars*

Even in early history, the ancient Chinese scholars already exhibited substantial expounding on the biological characteristics of Lingzhi. In *Lie Zi* (*Scholars*), the author observed that "Zhi grows above the rotten soil." In *Lun Heng* (*Speaking of Balance*), Wang Chong (27~about 97 A.D.) of the Eastern Han Dynasty pointed out that "Lingzhi grows from soils with balanced conditions." And Tao Hong-Jing (456~536 A.D., the Southern Dynasties) stated that "Zizhi, of the appearance of mushrooms, grows from the rotted tree trunks." It is evident that in the ancient times, it was already known that Lingzhi requires rotten soil or decayed wood to grow. *Bao Pu Zi* (*Magic Medicines*) edited by Ge Hon (284~364 A.D., the Eastern Jin Dynasty) described Lingzhi as organisms with red coral, white fat, lacquer black, peacock green, or golden yellow color and were bright and transparent like ice, in the sizes ranging from 5 kg to 1.5 kg. The color, appearance, and weight of the Lingzhi fruiting body were rather precisely recorded.

Furthermore, the mention of such things as "growing without flower are Zhi mushrooms" from *Li Ji Zhu Shu* (*Liji Notes*), "rootless plant of Lingzhi" from *ErYa Zhu Shu* (*Erya Notes*), "three crops a year of Lingzhi" and "all 6 kinds of Lingzhi can be harvested in June and August in a year" from the *Ben Cao Gang Mu* (*Compendium of Materia Medica*) show that the facts of these fungi are different from higher plants. The facts that Lingzhi have no roots, stem, and leaf differentiation, do not flower, and can be harvested several times in a year were known among the people working with Lingzhi in the early times.

The fact that Lingzhi can be used for medicinal purposes or as food was also recorded in Chinese history. For example, Wang Chong of the Eastern Han Dynasty stated in Lun Heng (*Speaking of Balance*), "Lingzhi produces 3 crops in a year. Including it in diet can result in longevity as it is god's food." The legendary Chinese doctor, Li Shi-Zhen (Fig. 1.2), pointed out that, "Lingzhi has been harvested around the year for the gods. It is safe for consumption and should be considered as a vegetable." Tao Hong-Jing believed that, "One may eat Lingzhi without a concern of

Fig. 1.2 Li Shi-Zhen (1518~1593 A.D., Minng Dynasty). (Reproduced with permission from Ref. [2])

dosage." On the other hand, Su Jing (599–674 A.D., Tang dynasty) thoughtfully indicated that, "Lingzhi is so rare. It is difficult to find them. How can anyone expect to have it continuously for a long period of time?" Therefore, it is apparent that the Chinese at the time already knew Lingzhi's health benefits and antiaging effects. However, the limited availability of the naturally grown Lingzhi prevented widespread applications.

Misinformation on Lingzhi exists in many Chinese medical literatures, including the classic *Shen Nong Ben Cao Jing* (*Shennong Materia Medica*). Criticisms by ancient scholars were found. Su Jing disagreed with the association of 5 Lingzhi with their growing areas: "It was found that the red Lingzhi was not limited to Huo mountain., the black not limited to Mt. Heng, the blue Lingzhi not limited to Mt. Tai, the white Lingzhi not limited to Hua mountain, and the yellow Lingzhi not limited to Song mountain." Su's observation is correct, as we know these varieties of wild Lingzhi can indeed be found in most parts of China. In the *Ben Cao Gang Mu* (*Compendium of Materia Medica*), Li Shi-Zhen reclassified Lingzhi according to the 5 colors and "5 Elements." His different viewpoints included that "Lingzhi of different colors logically taste different, but their taste may not necessarily be correlated to their color." More importantly, Li criticized the superstitious aspects associated with Lingzhi. Li also showed an accurate picture of Lingzhi in his book (Fig. 1.3) [2–6].

1.1.3 Overview of the History of Modern Research on Ganoderma *(Lingzhi)*

In the 1930s, Chinese scholar Deng Shuqun studied Ganoderma with modern scientific technology, and collected work in *Chinese fungi* (1964) [7]. Fungi taxonomists Zhao Jiding and Zhang Xiaoqing and Mao Xiaolan compiled the *Ganoderma*

Fig. 1.3 Illustration of Lingzhi in Ben Cao Gang Mu. (Reproduced with permission from Ref. [2])

(Lingzhi) in China (1981), the *New edition of China Ganoderma (Lingzhi)* (1989), and *The macrofungi in China* (2000) on the basis of the research work. These books detail the distribution, classification of *Ganoderma* in China, the biological characteristics, etc. [8–10].

Zhao JD et al. (1989) discussed the classification of "six Zhi" described in *Shennong Materia Medica* using a modern biological classification system. The results concluded that these "sixZhi" are not all Lingzhi and indicate Qingzhi (blue Lingzhi) could be a species similar to *Coriolus versicolor* or *Polystictus versicolor*; Huangzhi (yellow Lingzhi) is speculated to be the *Laetiporus sulphureus or Tyromyces sulphureus*; and Baizhi (white Lingzhi) as the possible fungus species is *Fomitopsis officinalis*. All these fungi do not belong to the *Ganoderma* genus [11].

In the 1950s, the Institute of Microbiology Chinese Academy of Sciences successfully achieved artificial cultivation of *Ganoderma lucidum*; then the other research organizations in Beijing, Wuhan, Shanghai, and Fujian also started research on the artificial cultivation of *Ganoderma lucidum* fruiting body, and cultivation of submerged fermentation mycelia and fermented liquid, and gradually extended to the country, for realizing the large-scale production for *Ganoderma lucidum* research and product development to provide ample raw materials [3].

In the 1970s, the Chinese medical scholars have carried out chemical and pharmacological studies on *Ganoderma lucidum* and preliminarily confirmed its medicinal value in clinical [12].

In the 1990s, the Chinese Pharmaceutical Association, China Edible Fungi association, and Beijing Medical University hosted the "First National Workshop of Ganoderma Research (1991)" and "First International Symposium on Ganoderma

Research (1994)" in Beijing, China. These academia meetings promoted modern research and industrial development of *Ganoderma*, and drove its application. The first edition of the *Modern research on Ganoderma (Lingzhi)* edited by Zhibin Lin, as a widely recognized academic monograph, was published in 1996. It was reprinted three more times in 2001, 2007, and 2015.

The *Pharmacopoeia of the Peoples Republic of China* (Part 1) (2000, 2005, 2010, and 2015 editions) listed the fruiting body of *Ganoderma lucidum* (Leyss. ex Fr.) Karst (Chi Zhi) (Fig. 1.4) and *Ganoderma sinensis* Zhao, Xu et Zhang (ZiZhi) (Fig. 1.5) as the legal traditional Chinese medicines. The efficacy of Lingzhi is "to

Fig. 1.4 *Ganoderma lucidum* (Leyss. ex Fr.) Karst (Chi Zhi) **a**. Growing *G. lucidum* in greenhouses; **b**. Growing immature fruiting body of *G. lucidum*; **c**. mature fruiting body of *G. lucidum*; **d**. The fruiting body of *G.lucidum* covered spore powder (Provide photos by Zhibin Lin)

Fig. 1.5 *Ganoderma sinensis* Zhao, Xu et Zhang (ZiZhi) (Reproduced with permission from Ref. [10])

Fig. 1.6 *Ganoderma tsugae* murr (Song shan Lingzhi) (Reproduced with permission from Ref. [10])

replenish *qi* and ease the mind, relieve cough and asthma" and its indications are "dizziness, insomnia, palpitation, shortness of breath, asthenic cough, asthma and loss appetite" [13].

Ganoderma tsugae murr (Song shan Lingzhi) (Fig. 1.6) has been approved by the Chinese government for health food production [3].

At present, *Ganoderma* (Lingzhi) preparation is not only used for clinical prevention and treatment of diseases but also used as health-care products for people, especially for the middle-aged and elderly population. In the future, the Chinese scholars should further scientifically study *Ganoderma* (Lingzhi) and reasonably apply and correctly evaluate it so as to make contributions to the health of human beings.

1.2 *Ganoderma* (Lingzhi) in the Traditional Chinese Culture

1.2.1 Archaeological Evidence of Using Ganoderma *(Lingzhi) in China*

In China, the earliest text recording of *Ganoderma* is the *Shennong Materia Medica*, which appeared more than 2100 years ago. Shennong, legend of the *God of Agriculture* in China, tasted a hundred herbs and encountered 70 poisonous ones in a day, giving the discovery process a vivid picture. He is said to be the first to collect and use medicinal plants in China. There is evidence for the use of medicinal plants dating back up to 60,000 years before present. However, owing to the lack of reliable archaeological evidence, it has been highly debated when prehistoric people began utilizing wild herbal medicine including *Ganoderma* (Lingzhi).

Recently, Yuan Y et al. (2018) examined the morphology of spores excavated from five *Ganoderma* samples (G1 – G5) in three archaeological sites located in Tianluoshan, Yuhangnanhu, and Qianjintadi that date back to the Neolithic era.

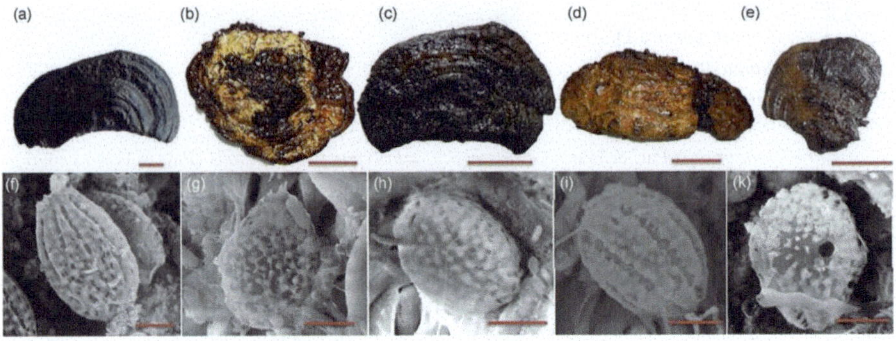

Fig. 1.7 The pileus surface and basidiospore morphological characteristics of ancient *Ganoderma* samples. The pileus surface of G1 (**a**), G2 (**b**), G3 (**c**), G4 (**d**), G5 (**e**); (**f**) the basidiopores of G1 are long ovoid, 5.4~7.3 μm long×3.5~4.8 μm width; (**g**) the basidiopores of G2 are long ovoid, 5.3 μm long×4.1 μm width; (**h**) the basidiopores of G3 are oblong, 5.3 μm long×3.1 μm width; (**i**) the basidiopores of G4 are oblong, 5.4~6.2 μm long× 3.5~4.6 μm width; (**k**) the basidiopores of G5 are peer-shaped, 4.9~5.7 μm long×3.0~4.3 μm width; the scale bar of (**a**) is 1.0 cm, the scales of (**b**) to (**e**) are 5.0 cm, the scales of (**f**) to (**k**) are 2 μm. (Reproduced with permission from Ref. [10])

Dating back to using ^{14}C isotope revealed that the use of G1 sample began about 6817 ± 44 years before present in the Hemudu society, G2 sample began about 5379 ± 59 years before present, and G5 sample began about 4508 ± 50 years before present in the Hemudu society and Liangchu society. The comparison of morphological characteristics of spores from the prehistoric samples and spores of the present-day, modern *Ganoderma* species confirmed that the G1–G5 samples belong to the genus *Ganoderma* (Fig. 1.7). The Hemudu society is one of the birthplaces of the Chinese civilization, and people had started using reed mats and planted tea and rice. The prehistoric *Ganoderma* was unearthed with the cultural relics, such as wood carvings, head ornaments, jade articles, and so on. It was speculated that the witch had been using *Ganoderma* at that time. The earliest dates for the lower Yangtze River areas Neolithic indicate that it expanded the archaeobotanical records of herbal medicine (*Ganoderma*) exploitation in China to 6800 years before present. With the formation of early agriculture, people continued the exploration and utilization of fungi with *Ganoderma* appearance. In the course of the history, the ancient Chinese people gradually documented its value and extolled its purpose. This opinion is related to the legendary events of "Xuanyuan gifted with *Ganoderma*" and "Shennong gathers *Ganoderma*" in the mythological era of China [14].

1.2.2 Lingzhi in Myths

More than 2000 years ago, there was already evidence of the Chinese worshipping Lingzhi. Myths associated with Lingzhi can be found in history.

In the *Book of Shan Hai Jing* (*Mountains and Seas*) of the Warring States Period (476–221 B.C.), Yaoji, young daughter of Emperor Yan, was mystified that she turned into the herb Yaocao (Grass of Yao) after she died. A poet from Chu, Song Yu engaged her in the fairy tale love story with a god. The myth eventually made Yaoji the origin of Lingzhi.

In the *Bai She Zhuan* (*Legend of the White Snake*), the heroine White Snake went alone to the Emei mountain to steal the celestial herb (i.e., Lingzhi) in order to save her husband's life. She overcame all sorts of hardships and finally moved the heart of the god who let her have the magical Lingzhi that revived her husband from the dead. The love story has become the subject of countless novels, dramas, movies, and posters in China (Fig. 1.8).

Ge Hong, in his *Legend of the Gods*, the beautiful goddess, Magu, pursued Taoism at Guyu mountain and lived on the Panlai Isle. She brewed the Lingzhi wine specifically for the Queen's birthday. This picture of Magu holding the wine, a child raising a birthday peach-shaped cake, an old man with a cup and a crane with Lingzhi in its mouth has become a popular folk art for birthday celebration with the wishes of fortune and longevity (Fig. 1.9) [2, 15].

1.2.3 Lingzhi in Ancient Poems

Lingzhi is the symbol of sanctity and goodness. It was a common motif in the Chinese literature in the past.

A famous poet from Chu, Qu Yuan described in his poem, *Nine Sons, Mountain Ghost*, a goddess longing for love. "Sanxiu" in the poem was another name for

Fig. 1.8 *Legend of the White Snake* (Steal the celestial herb). (Reproduced with permission from Ref. [2])

Fig. 1.9 Poster of Magu for birthday wishes. (Reproduced with permission from Ref. [1])

Lingzhi, which can be harvested more than once a year. The "Mountain Ghost" was the goddess who picked Lingzhi in Wu mountain.

When Emperor Hanwudi officiated the sacrificial ritual, 70 young boys and girls would chant *The Song of Lingzhi* with the music. The lyric carried the messages of wishing for auspiciousness, good fortune, and longevity.

In the Period of Three Kingdoms, poet Cao Zhi referred frequently to Lingzhi. For instance, in his well-known poem, *Lingzhi Pian* (*On Lingzhi*), he praised Lingzhi as creating the heaven and earth; the maroon herb growing along the bank of Luo river symbolized the prosperity of the nation and the glory of the god. In his poem *Luo Shen Fu* (*In Praise of the Goddess Luo*), the poet depicted how graceful and beautiful the Goddess Luo was while harvesting Lingzhi and his admiration of the goddess. In the *Fei Long Pian* (*Flying Dragon*), he told a story of his encounter in the misty Tai mountain with a Taoist monk on a white deer with a Lingzhi in his hand. It was from that man he learned the magical health benefits of Lingzhi.

In the *Chang Ge Xing* (*Singing Trip*) of a Han Dynasty Yuefu-style poem, a similar story was told. The author met a god-like man with short hair and long ears riding a white deer and was led to pick Lingzhi. At the legendary man's home, he was shown the tonic herb made from the health-improving, hair-color darkening, and life-prolonging effects of the red herb [1, 15].

1.2.4 "Catholicon" According to Taoism

The Lingzhi culture was greatly influenced by Taoism, the native religion in China. Taoism believes that living is most important and that human beings can be immortal by following the regimens and taking certain magical herbs. *Bao Pu Zi* written by Ge Hong presented the theory suggesting that a person could learn to become immortal. It even included stories of such occurrences by taking Lingzhi.

The ancient Taoist theory considered Lingzhi as the best among the catholicons, and by consuming Langzhi, one would never grow old or die. Therefore, Lingzhi acquired the names such as Shenzhi (heavenly herb) and Xiancao (magic grass) and became mystified. Most of the famous Taoists in history, including Ge Hong, Lu Xiu-Jing (406–477 A.D.), Tao Hong-Jing, and Sun Si-Miao (541–682 A.D.), saw the importance of Lingzhi studies. They influenced greatly in promoting the Lingzhi culture in China. In pursuing immortality, the Taoists enriched the knowledge on the herb and led to the evolution of the Taoist medical practice, which emphasizes health and well-being.

For their philosophy as well as a lack of scientific knowledge, the Taoists' understanding of Lingzhi was not only limited but also mostly superstitious. The term, "Zhi," used by them referred to many other kinds of fungi. It even included the mythical and imaginary herb. The religious connection was criticized by the medical profession in China and impeded the progress of Lingzhi's applications and true understanding [2, 3, 15].

1.2.5 Lingzhi Worship and Auspicious Symbol for the Chinese

Since the Han Dynasty, Confucian scholars gave Lingzhi the names of "fortune herb" or "fortune grass." They considered the circular lines on top of Lingzhi cap an auspicious symbol or fortunate halos. Gradually, the lucky charm characteristic of Lingzhi became a unique component in the Chinese culture.

For the Chinese, Lingzhi and its derivative, Ruyi (literally, as one wishes), have been the symbol for luck, fortune, longevity, peace, and prosperity. This is widely believed in and deeply affects the people and culture to date. Images of Lingzhi, and the "fortunate clouds" derived from it, can be found on palaces, temples, ancient buildings, clothing, embroidery, paintings, sculptures, china, and excavated archaeological relics. For example, they are on the pole in front of the Tiananmen Square, on the ceiling of the Qinian Hall at the Temple of Heaven, and on the royal passage leading to the Main Hall at the Forbidden City. There are carved Lingzhi bonsai on the fences at the Forbidden City, the Ancient Ministry of Education Building and the Confucius Temple, the graphic design on the base of the stone tablet at the Confucius Temple, as well as the wood carving of Lingzhi bonsi before the Sakyamuni statue in Yonghe Lamasery. These give the evidence of how the Chinese worshiped Lingzhi in the old times.

The silk painting by Emperor Qianlong of the Qing Dynasty (Fig. 1.10), a collection at the Palace Museum in Taipei, shows a vase with pine branches, camellia, and plum blossom. And on its side there are persimmon, lilies, and Lingzhi. This is a typical painting reflecting the wish for good luck and fortune.

To date, many Chinese idioms, such as "Ji Xiang Ru Yi" (good fortune and happiness), "Ci Fu Jia Xiang" (bestow blessing and happiness), "Zeng Tian Shou Kao"

Fig. 1.10 The silk (Ru Yi painting by Emperor Qianlong). (Reproduced with permission from Ref. [2])

(blessed with longevity), and "Guo Tai Ming An" (country is prosperous and people live in peace), continue to be used by the people, reflecting the ingrained Lingzhi culture [2].

1.3 Conclusion

Shennong Materia Medica and many other written records in the early Chinese history began to study, discuss, and report the scientific aspects of Lingzhi in respect to its categorization, habitat, bionomics, herbal nature, medication, etc. They have been frequently referred to in literature and used for further research and applications.

Ganoderma (Lingzhi) has been viewed as a magic herb as well as an auspicious symbol by the Chinese. It is, therefore, also known as "Ruizhi," "Shenzhi," and "Xiancao," meaning good fortune and mysterious power. Numerous myths and poems mentioning people's love, worship, and beliefs on Lingzhi can be found in the Chinese literature since ancient times.

Recent research on *Ganoderma* (Lingzhi) promotes the development of *Ganoderma* industrial application.

References

1. Anonymous (1997) Shen Nong Ben Cao Jing (Shennong Materia Medica), 1st edn. Liaoning Science and Technology Press, Shenyang, p 9
2. Lin ZB (ed) (2009) Lingzhi from mystery to science, 1st ed. Peking University Medical Press, Beijing, pp 1–19
3. Lin ZB (2015) Chapter 1. Historical data of Chinese Ganoderma (Lingzhi) research. In: Lin ZB (eds) Modern research on Ganoderma (Lingzhi), 4thed, Peking University Medical Press, Beijing, pp 1–7
4. Shizhen L (1978) Ben Cao Gang Mu (Compendium of Materia Medica), vols 3, 28. People's Medical Publishing House, Beijing, pp 1708–1712
5. Chen SY (1991) Textural research on a list of lost books on glossy Ganodermas in ancient China. China Hist Mat Sci Technol 12(3):70–80
6. Cao LJ, Zhu JP, Yuan B, Fang XY (2013) Studies of early stage on the efficacy of Ganoderma lucidum in China. Asia-Pasific Trad Med 9(5):67–69
7. Deng SQ (1964) Chinese fungi, 1st edn. Science Press, Beijing, pp 445–454
8. Zhao JD (1981) Chinese Ganoderma, 1st edn. Science Press, Beijing, pp 1–78
9. Zhao JD (1989) New edition of Chinese Ganoderma, 1st edn. Science Press, Beijing, pp 3–26
10. Mao XL (ed) (2000) The macrofungi in China, 1st edn. Henan Science and Technology Press, Zhengzhou, pp 483–504
11. Zhao JD, Zhang XQ (1989) Textual research on the classification of "six zhi" described by Chinese ancient books. Microbiol Bull (3):180–181
12. Lin ZB (1979) Current situation of Ganoderma research in China. Acta Pharm Sin 14(3):183–192
13. China Pharmacopoeia Committee (2015) The pharmacopoeia of the Peoples Republic of China (Part 1), 2015th ed. China Medical Science and Technology Press, Beijing, pp 188
14. Yuan Y, Wang Y, Sun G, Wang Y, Cao L, Shen Y, Yuan B, Han D, Huang L (2018) Archaeological evidence suggests earlier use of *Ganoderma* in Neolithic China. Sci Bull 63(13):1180–1188
15. Chen SY, Chen QW (2003) Fungal anthropology and Ganoderma culture. J Hubei Agric Coll 23(6):426–433

Chapter 2
Classification, Biological Characteristics and Cultivations of *Ganoderma*

Zhuo Du, Cai-Hong Dong, Ke Wang, and Yi-Jian Yao

Abstract Species of *Ganoderma* (Ling-zhi) have been widely researched and cultivated due to their highly prized medicinal value, which is famous as a traditional Chinese medicine. The aims of this chapter are to (1) review the historical taxonomy of the family Ganodermataceae, (2) provide an account of the genera and species of *Ganoderma* together with the distributions and habitats, (3) evaluate morphological features and phylogenetic methods to define the genera and species and (4) present two commonly used cultivated methods (wood-log cultivation and substitute cultivation) for *Ganoderma*.

Keywords *Ganoderma* · Ling-zhi · taxonomy · biological characteristics · wood-log cultivation · substitute cultivation

2.1 Introduction

2.1.1 Definition and Circumscription of Ling-zhi

Ganoderma P. Karst. is a cosmopolitan genus of wood-decaying polypore fungi, typified by *Ganoderma lucidum* (Curtis) P. Karst. [1] and usually recognised by ellipsoid to ovoid, double-walled and truncate basidiospores [1–3]. *Ganoderma* species have a worldwide distribution in green ecosystems both in tropical and temperate geographical regions and grow as a facultative parasite that can live as saprobes on rotting stumps and roots.

Ling-zhi, the widely cultivated *Ganoderma* fungus, is famous for its medicinal values well documented in the Chinese literature which can be dated back nearly 2000 years to the Shen Nong Materia Medica (102–200 AD) [4]. It symbolises hap-

Z. Du · C.-H. Dong · K. Wang · Y.-J. Yao (✉)
State Key Laboratory of Mycology, Institute of Microbiology, Chinese Academy of Sciences, Beijing, China
e-mail: yaoyj@im.ac.cn

Z. Lin, B. Yang (eds.), *Ganoderma and Health*, Advances in Experimental Medicine and Biology 1181, https://doi.org/10.1007/978-981-13-9867-4_2

piness, good fortune, good health and even immortality in Chinese traditional culture [5]. There are thousands of publications relating to this fungus and it is now commercially cultivated on a large scale.

2.1.2 *Classification Position of* Ganoderma

Ganoderma is a genus of polypore Fungi in the kingdom fungi, phylum Basidiomycota, class Agaricomycetes, order Polyporales and family Ganodermataceae. There are more than 200 species in the genus, many of them from tropical regions. Because of their genetic diversity, use in traditional Asian medicines and potential in bioremediation, *Ganoderma* is an important genus economically. *Ganoderma* can be differentiated from other polypores because they have a double-walled basidiospore. They are sometimes called shelf mushrooms or bracket fungi.

2.1.3 *Genera of the Family Ganodermataceae*

According to the international fungal name database Index Fungorum (http://www.indexfungorum.org/, accessed 30th March, 2019), there are seven accepted genera and another eight synonyms in the family Ganodermataceae, as listed in Table 2.1.

Table 2.1 Genera in the Ganodermataceae

Genus	Synonyms
Amauroderma Murrill	=*Lazulinospora* Burds. & M.J. Larsen
	=*Magoderna* Steyaert
	=*Whitfordia* Murrill
Foraminispora Robledo, Costa-Rezende & Drechsler-Santos	
Furtadoa[a] Costa-Rezende, Robledo & Drechsler-Santos	
Ganoderma P. Karst.	=*Dendrophagus* Murrill
	=*Elfvingia* P. Karst.
	=*Friesia* Lázaro Ibiza
	=*Tomophagus* Murrill
	=*Trachyderma* (Imazeki) Imazeki
Haddowia Steyaert	
Humphreya Steyaert	
Polyporopsis Audet	

[a]Note: this genus name was regarded as an illegitimate name and need more research

2.1.4 Important Species of Ganoderma

Medicinal *Ganoderma* **Species** Species of *Ganoderma* are widely researched because of their highly prized medicinal value. *Ganoderma amboinense* (Lam.) Pat., *G. applanatum* (Pers.) Pat., *G. lucidum* (Curtis) P. Karst. (a misapplied name to Ling-zhi widely cultivated in China, the correct name is G. sichuanense J.D. Zhao & X.Q. Zhang.), *G. tsugae* Murrill are regarded as having an anticancer property [6–8]. *Ganoderma applanatum, G. atrum* J.D. Zhao, L.W. Hsu & X.Q. Zhang and *G. lucidum* are reported to have an antidiabetic effect [9–11]. *G. applanatum*; *G. atrum*; *G. capense* (Lloyd) Teng; *G. colossus* (Fr.) C.F. Baker; *G. lucidum*; *G. sichuanense*; *G. sinense* J.D. Zhao, L.W. Hsu & X.Q. Zhang and *G. tsugae* are utilised for their anti-inflammatory properties [12–20]. In addition, *G. lucidum* and *G. pfeifferi* Bres. are good sources of antioxidant compounds [21–25]. In some cardiovascular problems, the presence of polysaccharides in *G. lucidum* specifically have a therapeutic effect [26].

Ganoderma **Products** Two species, Ling-zhi (*G. lucidum*) and Zi-zhi (*G. sinense*), are recorded as legal medicinal fungi in *Pharmacopoeia of the People's Republic of China* 2015 edition [27]. Three species, Ling-zhi, Zi-zhi and *G. tsugae*, are recorded in "the list of fungal species that can be used in health food" authorised by the Chinese government [9]. *Ganoderma lucidum* products, extracted from different parts of its fruit bodies, mycelia or spores, are sold in the form of coffee, powder, tea, dietary supplements, spore products, drinks, syrups, toothpastes, soaps and lotions. It has been widely commercialised as food and drug supplements which enhance the body's immune system and improve metabolic functions [28–30].

Except being used as the material of functional food in daily life, *Ganoderma* is also a part of several cosmetics like skin lightening, which are produced mainly in China, Korea, the USA and some other Asian and European countries [31]. Tyrosinase enzyme is a key enzyme in melanin formation and *G. lucidum* has shown tyrosinase inhibition activity [32]. Hence, many facial mask cosmetics in the market contain *Ganoderma* extracts which help in skin whitening [33].

Ganoderma lucidum was used as bonsai products to decorate gardens, ornaments and many other art products [34]. *Ganoderma sinense* also has high ornamental value to make bonsai because of its hard texture and good colour [35].

Cultivation of *Ganoderma* **Species** Cultivation of Ling-zhi can be traced back to 1959 when fruit bodies of the fungus were successfully grown in the Institute of Microbiology, Chinese Academy of Sciences in Beijing, supported by a collection of basidiomata preserved in the Fungarium (HMAS) of Chinese Academy of Sciences. A variety of cultivated or useful species of *Ganoderma* were recorded in literatures, for example *Ganoderma amboinense*; *G. applanatum*; *G. boninense* Pat.; *G. brownie* (Murrill) Gilb.; *G. capense* (Lloyd) Teng; *G. duropora* Lloyd; *G. gibbosum* (Cooke) Pat.; *G. guinanense* J.D. Zhao & X.Q. Zhang; *G. lobatum* (Cooke) G.F. Atk.; *G. lucidum*; *G. resinaceum* Boud.; *G. sinense*; *G. tenue* J.D. Zhao, L.W. Hsu & X.Q. Zhang; *G. tropicum* (Jungh.) Bres. and *G. tsugae* [36].

2.2 Taxonomy and Classification of *Ganoderma*

2.2.1 Taxonomic History

2.2.1.1 Worldwide

The genus *Ganoderma* was established by Karsten [1] for the laccate and stipitate white rot fungus *Polyporus lucidus* (Curtis) Fr. Later, Patouillard [37] amended it to include all polypores having double-walled basidiospores and divided the genus into two sections, sect. *Ganoderma* and sect. *Amauroderma*. The former was created to separate species characterised by subspherical or spherical basidiospores with the wall uniformly thickened from those having basidiospores truncated at the apex. This classification system was followed by contemporary taxonomist [38, 39]. Karsten established the genus *Elfvingia* P. Karst. for nonlaccate *Ganoderma* species. However, *Elfvingia* was emended to include all species lacking pilocystidia and reduced to subgeneric rank by Imazeki [40], which was considered to be a synonym of *Ganoderma* [41]. Murrill raised *Amauroderma* Murrill to generic rank [42]. Then the genus *Tomophagus* Murrill for *P. colossus* Fr. was established basing on the unique spongy basidiocarp [43, 44]. But the genus was also considered as a synonym of *Ganoderma* by Furtado [41], Steyaert [45] and Ryvarden [46]. Donk broke the previous concepts and upgraded *Ganoderma* to Ganodermataceae, including *Ganoderma* and *Amauroderma* [47]. Later, the genus *Ganoderma* was divided into three subgenera. Laccate species were classified as subgen. *Euganoderma*; the perennial, nonlaccate species with dark brown context were classified as subgen. *Elfvingia*; and the subgen. *Trachyderma* was established to contain annual, nonlaccate, white context *G. tsunodae* (Yasuda ex Lloyd) Sacc. like species [40]. In 1943, the concept of Ganodermataceae was not accepted, and Ganodermeae was established under Polyporaceae, including *Amauroderma, Elfvingia* and *Ganoderma*. *G. tsunodae* was moved to genus *Elfvingia* [48]. Imazeki upgraded subgen. *Trachyderma* to genus *Trachyderma* (Imazeki) Imazeki [49]. Steyaert proposed three new genera, *Haddowia* Steyaert, *Humphreya* Steyaert and *Magoderna* Steyaert, to solve the mixture of *Ganoderma* and *Amauroderma* species [45]. Later, *Ganoderma* was divided into four subgenera and two sections on the basis of the cutis anatomy [50], including *Ganoderma* (sect. *Ganoderma* and sect. *Characoderma*), *Elfvingia*, *Anamixoderma* and *Plecoderma*. In the taxonomy system of Zhao et al., Ganodermataceae was accepted, and on the basis of context colour, subgen. *Ganoderma* was segregated into sect. *Ganoderma* and *Phaeonema* [51]. Jülich established Ganodermatales consisting of Ganodermataceae and Haddowiaceae [52]. Ganodermataceae was subdivided into *Amauroderma*, *Ganoderma* and *Humphreya*. Corner supported the five genera in the taxonomy system of Steyaert [45] and noted intermediacy between the described types of cutis, arguing that alleged distinctions break down in practice [53]. Ryvarden described the morphological characteristics of *Ganoderma* in detail and pointed out the differences between *Amauroderma* and *Ganoderma* [54]. Another important morphological feature was discovered: the presence of chlamydospores is an important characteristic for species identification and it is also corroborated by molecular phylogenetic studies. Only a few *Ganoderma* species, such as *G. colossus*, *G. subamboinense* Henn. and *G. weberianum* (Bres. & Henn. ex Sacc.) Steyaert, contain chlamydospores in their basidiomata [55, 56]. In the *Dictionary of the Fungi*, four genera were recorded [57]. In recent years, three genera had been established. The new genus *Polyporopsis* Audet. was introduced by Costa-Rezende et al. to accommodate *Albatrellus mexicanus* Laferr. & Gilb [58]. In 2017, the presence of perforation in the exosporium with holes that were connected with hollow columns of the endosporium was considered as a synapomorphy in *Foraminispora* Robledo, Costa-Rezende & Drechsler-Santos, a new genus proposed to accommodate *Porothelium rugosum* Berk. [3]. *Furtadoa* Costa-Rezende, Robledo & Drechsler-Santos was established as a new genus, which was characterised by having stipitate basidiomata, soft when fresh, dull pilear surface, pale context, adimitic hyphal system, with a monomitic context, composed of both clamped and simple-septate generative hyphae [3].

According to Index Fungorum (http://www.indexfungorum.org/, 2019), there are seven accepted genera and another eight synonyms in the family Ganodermataceae, and 499 species epithets were listed in the genus *Ganoderma*.

2.2.1.2 China

Ling-zhi, a widely cultivated fungus in China, has a long history in traditional Chinese medicine. The history of Ling-zhi can be traced to the ancient book named *Shen Nong Materia Medica*, which recorded several groups of *Ganoderma*, that is Bai-zhi, Chi-zhi, Hei-zhi, Huang-zhi, Qing-zhi and Zi-zhi [59]. The earliest national pharmacopoeia – *Tang Materia Medica* – continued the records of the six groups of Ling-zhi [60]. In the *Compendium of Materia Medica* of the Ming Dynasty, Ling-zhi belonged to "vegetables" and had a great impact on Chinese culture owing to its medicinal benefits to human health [61]. Therefore, Ling-zhi has been regarded as a rare medicine in China since ancient times.

According to Tai, Patouillard first reported *G. lucidum* in China in 1907 based on specimens collected from Guizhou Province collected in 1896 by Cavalerie [62]. Teng reported four species and one variety of *Ganoderma* and provided detailed morphological descriptions [63]. Teng added the records of 26 species of *Ganoderma* and *Amauroderma* [64]. Tai summarised 38 reported species in China [62]. Zhao et al. accepted that Ganodermoideae contained *Amauroderma* and *Ganoderma*, confirmed 53 species distributed in China and divided *Ganoderma* into sect. *Ganoderma* and *Phaeonema* [51]. Later, Zhao reported 86 species belonging to four genera: *Amauroderma*, *Ganoderma*, *Haddowia* and *Humphreya* [63]. Zhao and Zhang [64] reported 98 species in Ganodermataceae, of which 76 belonged to *Ganoderma*, including 48 new species. The genus *Ganoderma* was separated into three subgenera: *Elfvingia*, *Ganoderma* and *Trachyderma* [64]. Records of Ganodermataceae increased to 103 species in China [65]. Wang et al. analysed both morphology and molecular data and proposed that the identity of the Chinese "*G. lucidum*" (Ling-zhi) should be considered as a synonym of *G. sichuanense*, a species originally described from Sichuan Province, China [66]. Cao et al. proposed a new species *G. lingzhi* S.H. Wu, Y. Cao & Y.C. Dai for commercially cultivated "*G. lucidum*" [67]. In order to interpret the species concept of the Chinese "*G. lucidum*", Yao et al. designated an epitype of *G. sichuanense* [68], which has been accepted in other publications [69–71].

2.2.2 *Number of Species and Infraspecific Taxa Described in* Ganoderma

According to the international fungal name database Index Fungorum (http://www.indexfungorum.org/, accessed 30th March, 2019), there are 213 accepted species and infraspecific taxa in the genus *Ganoderma*, as listed in Table 2.2.

2.2.3 *Infragenic Classification*

The genus *Ganoderma* was named by Karsten in 1881 [1]. Members of the family Ganodermataceae were traditionally considered to be difficult to classify because of the lack of reliable morphological characteristics, the overabundance of synonyms and the widespread misuse of names [72, 73].

Imazeki was the first to divide the genus *Ganoderma* into several subgenera [40]. Species with laccate surface belonged to subgen. *Euganoderma*, perennial and no laccate species belonged to subgen. *Elfvingia* and the other annual species with white context belonged to subgen. *Trachyderma* [40].

Steyaert divided the genus into four subgenera and two sections according to the structure of crustification [50]. Subgenus *Ganoderma* was subdivided into two sections: *Ganoderma* and *Characoderma*. Subgen. *Elfvingia* was clearly distinct in having a cutis anatomy of the trichoderm type. Subgen. *Anamixoderma* was distinct by cutis with hyaline and brown hyphae loosely intermixed, totally embedded in melanoid substances. Subgen. *Plecoderma* was distinct by cutis of hyaline hyphae densely entwined and impregnated by melanoid substances, forming a layer sharply distinct from the context [50].

Zhao amended the under-genus taxa completely. Subgenus *Ganoderma* contained laccate species and was divided into two sections – sect. *Ganoderma* with light-coloured context and sect. *Phaeonema* with dark-coloured context. The no laccate species were divided into subgen. *Elfvingia* and subgen. *Trachyderma*. Subgen. *Elfvingia* contained dark-coloured species and subgen. *Trachyderma* contained light-coloured species [74].

Table 2.2 Species and infraspecific taxa described in *Ganoderma*

Current names	Synonyms
Ganoderma adspersum (Schulzer) Donk	=*Polyporus adspersus* Schulzer
	=*Ganoderma europaeum* Steyaert
Ganoderma africanum (Lloyd) Doidge	=*Polyporus africanus* Lloyd
Ganoderma ahmadii Steyaert	
Ganoderma albomarginatum S.C. He	
Ganoderma alluaudii Pat. & Har.	
Ganoderma amazonense Weir	
Ganoderma angustisporum J.H. Xing, B.K. Cui & Y.C. Dai	
Ganoderma applanatum (Pers.) Pat.	=*Ganoderma leucophaeum* (Mont.) Pat.
	=*Ganoderma applanatum* var. *philippinense* C.J. Humphrey & Leus-Palo
	=*Fomes applanatus* (Pers.) Fr.
	=*Boletus applanatus* Pers.
	=*Fomes megaloma* (Lév.) Cooke
	=*Polyporus leucophaeus* Mont.
	=*Polyporus applanatus* (Pers.) Wallr.
	=*Ungularia subganodermica* Lázaro Ibiza
	=*Fomes vegetus* (Fr.) Fr.
	=*Fomes incrassatus* (Berk.) Cooke
	=*Polyporus merismoides* Corda
	=*Polyporus subganodermicus* (Lázaro Ibiza) Sacc. & Trotter
	=*Polyporus megaloma* Lév.
	=*Fomes gelsicola* Berl.
	=*Fomes leucophaeus* (Mont.) Cooke
	=*Fomes stevenii* (Lév.) P. Karst.
	=*Ganoderma incrassatum* (Berk.) Bres.
	=*Ganoderma applanatum* var. *laevisporum* C.J. Humphrey & Leus-Palo
	=*Fomes longoporus* Lloyd
	=*Fomes vegetus* var. *leucostratus* Yamano
	=*Fomes applanatus* var. *leucophaeus* (Mont.) Cleland & Cheel
	=*Fomes applanatus* var. *applanatus* (Pers.) Fr.
	=*Polyporus applanatus* f. *applanatus* (Pers.) Wallr.
	=*Ganoderma applanatum* subsp. *applanatum* (Pers.) Pat.
	=*Fomes applanatus* f. *coralloides* Glaser
	=*Ganoderma applanatum* f. *macrosporum* C.J. Humphrey & Leus-Palo

(continued)

Table 2.2 (continued)

Current names	Synonyms
	=*Ganoderma megaloma* (Lév.) Bres.
	=*Phaeoporus applanatus* (Pers.) J. Schröt.
	=*Placodes applanatus* (Pers.) Quél.
	=*Polyporus vegetus* Fr.
	=*Ganoderma applanatum* var. *applanatum* (Pers.) Pat.
	=*Ganoderma incrassatum* f. *incrassatum* (Berk.) Bres.
	=*Fomes vegetus* var. *vegetus* (Fr.) Fr.
	=*Fomes applanatus* f. *applanatus* (Pers.) Fr.
	=*Elfvingia applanata* (Pers.) P. Karst.
	=*Ganoderma lipsiense*
	=*Ganoderma applanatum* f. *applanatum* (Pers.) Pat.
	=*Fomes applanatus* f. *leucophaeus* (Mont.) Lloyd
	=*Elfvingia megaloma* (Lév.) Murrill
	=*Ganoderma applanatum* var. *vegetum* (Fr.) Rea
	=*Friesia applanata* (Pers.) Lázaro Ibiza
	=*Friesia vegeta* (Fr.) Lázaro Ibiza
	=*Ganoderma flabelliforme* Murrill
	=*Ganoderma gelsicola* (Berl.) Sacc.
	=*Scindalma gelsicola* (Berl.) Kuntze
	=*Scindalma incrassatum* (Berk.) Kuntze
	=*Scindalma leucophaeum* (Mont.) Kuntze
	=*Scindalma megaloma* (Lév.) Kuntze
	=*Scindalma stevenii* (Lév.) Kuntze
	=*Scindalma vegetum* (Fr.) Kuntze
	=*Polyporus concentricus* Cooke
	=*Polyporus incrassatus* Berk.
	=*Polyporus stevenii* Lév.
	=*Polyporus applanatus* f. *stratosus* Sacc.
	=*Boletus fomentarius* var. *applanatus* (Pers.) Pers.
	=*Ganoderma incrassatum* f. *substipitatum* Bres.
	=*Fomes australis* var. *applanatus* (Pers.) Sacc.
	=*Placodes vegetus* (Fr.) Quél.
Ganoderma aridicola J.H. Xing & B.K. Cui	
Ganoderma asperulatum (Murrill) Sacc. & Trotter	=*Amauroderma asperulatum* Murrill

(continued)

Table 2.2 (continued)

Current names	Synonyms
Ganoderma atrum J.D. Zhao, L.W. Hsu & X.Q. Zhang	
Ganoderma aureolum Steyaert	
Ganoderma auriscalpioides Henn.	
Ganoderma australe (Fr.) Pat.	=*Ganoderma tornatum* (Pers.) Bres.
	=*Ganoderma applanatum* var. *tornatum* (Pers.) C.J. Humphrey & Leus-Palo
	=*Fomes pseudoaustralis* Lloyd
	=*Fomes australis* subsp. *arculatus* (Bres.) Sacc.
	=*Fomes australis* (Fr.) Cooke
	=*Fomes scansilis* (Berk.) Cooke
	=*Fomes arculatus* (Bres.) Sacc.
	=*Fomes annularis* Lloyd
	=*Polyporus australis* Fr.
	=*Fomes undatus* Lázaro Ibiza
	=*Polyporus scansilis* Berk.
	=*Polyporus tornatus* Pers.
	=*Ganoderma arcuatum* (Bres.) Mussat
	=*Fomes australis* var. *oroflavus* (Lloyd) G. Cunn.
	=*Ganoderma annulare* (Lloyd) Boedijn
	=*Fomes applanatus* var. *australis* (Fr.) Cleland & Cheel
	=*Fomes applanatus* var. *oroflavus* (Lloyd) Cleland & Cheel
	=*Elfvingia australis* (Fr.) G. Cunn.
	=*Ganoderma oroflavum* (Lloyd) C.J. Humphrey
	=*Elfvingia tornata* (Pers.) Murrill
	=*Fomes oroflavus* Lloyd
	=*Fomes australis* subsp. *australis* (Fr.) Cooke
	=*Polyporus australis* var. *australis* Fr.
	=*Fomes australis* var. *australis* (Fr.) Cooke
	=*Ganoderma tornatum* var. *tornatum* (Pers.) Bres.
	=*Ganoderma applanatum* f. *australe* (Fr.) Bourdot & Galzin
	=*Fomes polyzonus* Lloyd
	=*Ganoderma australe* f. *arculatum* Bres.
	=*Scindalma arculatum* (Bres.) Kuntze
	=*Scindalma scansile* (Berk.) Kuntze

(continued)

Table 2.2 (continued)

Current names	Synonyms
	=*Scindalma tornatum* (Pers.) Kuntze
	=*Ganoderma australe* f. *australe* (Fr.) Pat.
	=*Ganoderma applanatum* subsp. *australe* (Fr.) Bourdot & Galzin
	=*Fomes nigrolaccatus*
	=*Placodes australis* (Fr.) Quél.
Ganoderma austroafricanum M.P.A. Coetzee, M.J. Wingf., Marinc. & Blanchette	
Ganoderma austrofujianense J.D. Zhao, L.W. Hsu & X.Q. Zhang	
Ganoderma barretoi Torrend	
Ganoderma baudonii Steyaert	
Ganoderma baumii Pilát	
Ganoderma bawanglingense J.D. Zhao & X.Q. Zhang	
Ganoderma bicharacteristicum X.Q. Zhang	
Ganoderma bilobum Bres.	
Ganoderma brownii (Murrill) Gilb.	=*Elfvingia brownii* Murrill
	=*Fomes brownii* (Murrill) Murrill
Ganoderma bruggemanii Steyaert	
Ganoderma calidophilum J.D. Zhao, L.W. Hsu & X.Q. Zhang	
Ganoderma cantharelloideum M.H. Liu	
Ganoderma capense (Lloyd) Teng	=*Polyporus capensis* Lloyd
Ganoderma carnosum Pat.	=*Ganoderma atkinsonii* H. Jahn, Kotl. & Pouzar
	=*Fomes carnosus* (Pat.) Sacc.
	=*Ganoderma valesiacum*
	=*Scindalma carnosum* (Pat.) Kuntze
Ganoderma carocalcareum Douanla-Meli	
Ganoderma casuarinicola J.H. Xing, B.K. Cui & Y.C. Dai	
Ganoderma cehengense X.L. Wu	
Ganoderma cervinum (Bres.) Sacc.	=*Amauroderma cervinum* Bres.
Ganoderma chalceum (Cooke) Steyaert	=*Ganoderma chalceum* var. *lobulatum* Corner
	=*Ganoderma chalceum* var. *pleiotrichum* Corner
	=*Ganoderma chalceum* var. *latiporum* Corner
	=*Ganoderma balabacense* Murrill
	=*Ganoderma cacainum* Bres.
	=*Irpex albofuscus* (Pat.) Sacc. & Trotter

(continued)

Table 2.2 (continued)

Current names	Synonyms
	=*Polyporus albofuscus* (Pat.) Trotter
	=*Coriolus albofuscus* Pat.
	=*Ganoderma hollidayi* Steyaert
	=*Ganoderma chalceum* var. *chalceum* (Cooke) Steyaert
	=*Polyporus chalceus* Cooke
Ganoderma chenghaiense J.D. Zhao	
Ganoderma chilense (Fr.) Pat.	=*Fomes chilensis* (Fr.) Sacc.
	=*Polyporus chilensis* Fr.
	=*Scindalma chilense* (Fr.) Kuntze
Ganoderma chiungchungense X.L. Wu	
Ganoderma chonoides Steyaert	
Ganoderma citriporum Ryvarden & Iturr.	
Ganoderma cochlear (Blume & T. Nees) Merr.	=*Fomes amboinensis* var. *pisachapani* (T. Nees) Sacc.
	=*Polyporus cochlear* Blume & T. Nees
	=*Polyporus pisachapani* T. Nees
	=*Ganoderma cochlear* f. *pisachapani* (T. Nees) Overeem
	=*Ganoderma cochlear* f. *cochlear* (Blume & T. Nees) Merr.
Ganoderma colossus (Fr.) C.F. Baker	=*Thermophymatospora fibuligera* Udagawa, Awao & Abdullah
	=*Polyporus hollandii* Massee
	=*Polyporus colossus* Fr.
	=*Tomophagus colossus* (Fr.) Murrill
	=*Dendrophagus colossus* (Fr.) Murrill
Ganoderma comorense (Henn.) Sacc. & Trotter	=*Fomes comorensis* Henn.
Ganoderma concinnum Ryvarden	
Ganoderma corrugatum Steyaert	
Ganoderma crebrostriatum J.D. Zhao & L.W. Hsu	
Ganoderma cupreopodium X.L. Wu & X.Q. Zhang	
Ganoderma cupulatiprocerum X.L. Wu & X.Q. Zhang	
Ganoderma curranii Murrill	=*Ganoderma curranii* var. *dimiticum* Corner
	=*Ganoderma curranii* var. *curranii* Murrill
Ganoderma curtisii (Berk.) Murrill	=*Fomes curtisii* (Berk.) Cooke
	=*Polyporus curtisii* Berk.
	=*Scindalma curtisii* (Berk.) Kuntze
Ganoderma dahlii (Henn.) Aoshima	=*Fomes dahlii* Henn.
Ganoderma daiqingshanense J.D. Zhao	

(continued)

Table 2.2 (continued)

Current names	Synonyms
Ganoderma dejongii Steyaert	
Ganoderma destructans M.P.A. Coetzee, Marinc. & M.J. Wingf.	
Ganoderma diaoluoshanense J.D. Zhao & X.Q. Zhang	
Ganoderma dimidiatum (Thunb.) V. Papp	=*Boletus dimidiatus* Thunb.
Ganoderma donkii Steyaert	
Ganoderma dorsale (Lloyd) Torrend	=*Polyporus dorsalis* Lloyd
Ganoderma dubio-cochlear (Lloyd) Sacc. & trotter	=*Polyporus dubio-cochlear* Lloyd
Ganoderma dussii Pat.	=*Fomes dussii* (Pat.) Sacc.
Ganoderma ecuadorense A. Salazar, C.W. Barnes & Ordoñez	
Ganoderma elegantum Ryvarden	
Ganoderma endochrum Steyaert	
Ganoderma enigmaticum M.P.A. Coetzee, Marinc. & M.J. Wingf.	
Ganoderma fasciatum Har. & Pat.	
Ganoderma fassii Steyaert	=*Ganoderma fassii* var. *dendroides* Steyaert
	=*Ganoderma fassii* var. *guineense* Steyaert
	=*Ganoderma fassii* var. *ipamuense* Steyaert
	=*Ganoderma fassii* var. *luteobadium* Steyaert
	=*Ganoderma fassii* var. *podagrum* Steyaert
	=*Ganoderma fassii* var. *ruwenzoriense* Steyaert
	=*Ganoderma fassii* var. *fassii* Steyaert
Ganoderma fassioides Steyaert	
Ganoderma fici Pat.	=*Fomes fici* (Pat.) Sacc.
Ganoderma flexipes Pat.	=*Polyporus flexipes* (Pat.) Lloyd
	=*Fomes flexipes* (Pat.) Sacc. & Traverso
Ganoderma formosanum T.T. Chang & T. Chen	
Ganoderma formosissimum (Speg.) Speg.	=*Fomes formosissimus* (Speg.) Sacc.
	=*Scindalma formosissimum* (Speg.) Kuntze
	=*Polyporus formosissimus* Speg.
Ganoderma frondosum Pat.	
Ganoderma fuscum Steyaert	
Ganoderma ghesquierei Steyaert	
Ganoderma gilletii Steyaert	
Ganoderma guianense Decock & Ryvarden	
Ganoderma guinanense J.D. Zhao & X.Q. Zhang	
Ganoderma guizhouense S.C. He	
Ganoderma hainanense J.D. Zhao, L.W. Hsu & X.Q. Zhang	

(continued)

Table 2.2 (continued)

Current names	Synonyms
Ganoderma hinnuleum Steyaert	
Ganoderma hoehnelianum Bres.	
Ganoderma hoploides Steyaert	
Ganoderma hypoxanthum (Bres.) C.J. Humphrey	=*Polyporus hypoxanthus* Bres.
Ganoderma impolitum Corner	
Ganoderma incrustatum (Fr.) Bres.	=*Ganoderma incrustatum* var. *bilobum* Bres.
	=*Fomes incrustatus* (Fr.) Sacc.
	=*Ganoderma incrustatum* var. *incrustatum* (Fr.) Bres.
	=*Polyporus incrustatus* Fr.
Ganoderma infulgens (Torrend) Sacc. & trotter	=*Polyporus infulgens* Torrend
	=*Amauroderma infulgens* (Torrend) Torrend
Ganoderma jianfenglingense X.L. Wu	
Ganoderma koningsbergii (Lloyd) Teng	=*Fomes koningsbergii* Lloyd
Ganoderma kosteri Steyaert	
Ganoderma kunmingense J.D. Zhao	
Ganoderma lamaoense Steyaert	
Ganoderma leucocontextum T.H. Li, W.Q. Deng, Sheng H. Wu, Dong M. Wang & H.P. Hu	
Ganoderma leucocreas Pat. & Har.	=*Polyporus leucocreas* (Pat. & Har.) Lloyd
Ganoderma leytense Steyaert	
Ganoderma lignosum Pat.	
Ganoderma linhartii (Kalchbr.) Z. Igmándy	=*Polyporus linhartii* Kalchbr.
Ganoderma lionnetii Rolland	=*Fomes lionnetii* (Rolland) Sacc. & D. Sacc.
	=*Elfvingia lionnetii* (Rolland) Murrill
Ganoderma lobatoideum Steyaert	
Ganoderma lobatum (Cooke) G.F. Atk.	=*Fomes lobatus* Cooke
	=*Polyporus lobatus* Schwein.
	=*Elfvingia lobata* (Cooke) Murrill
	=*Scindalma lobatum* (Cooke) Kuntze
Ganoderma lobenense Tonjock & Mih	
Ganoderma longistipitatum Ryvarden	
Ganoderma lucidum (Curtis) P. Karst. (misapplied to Chinese Ling-zhi)	=*Ganoderma ostreatum* Lázaro Ibiza
	=*Ganoderma lucidum* var. *nicotianae* (Inglese) Sacc. & Trotter
	=*Ganoderma lucidum* var. *resinosum* Pat.
	=*Fomes lucidus* var. *resinosus* Pat. ex Rick
	=*Polyporus laccatus* (Timm) Pers.
	=*Fomes lucidus* (Curtis) Sacc.
	=*Fomes lucidus* var. *badius* (Pat.) Sacc.
	=*Fomes resinaceus* var. *martellii* (Bres.) Sacc.

(continued)

Table 2.2 (continued)

Current names	Synonyms
	=*Fomes lucidus* var. *nicotianae* (Inglese) Sacc. & Traverso
	=*Polyporus lucidus* var. *exquisitus* Kalchbr.
	=*Polyporites lucidus* (Leyss.) Mesch.
	=*Fomes japonicus* (Fr.) Sacc.
	=*Boletus flabelliformis* Leyss.
	=*Ganoderma japonicum* (Fr.) Sawada
	=*Polyporus japonicus* Fr.
	=*Ganoderma lucidum* f. *annulatum* Torrend
	=*Ganoderma lucidum* f. *hemisphaericum* Torrend
	=*Ganoderma lucidum* f. *rubellum* Torrend
	=*Ganoderma lucidum* var. *resinaceum* Maire
	=*Ganoderma lucidum* var. *typicum* Maire
	=*Boletus lucidus* Curtis
	=*Ganoderma mongolicum* Pilát
	=*Fomes lucidus* var. *exquisitus* (Kalchbr.) F.M. Bailey
	=*Polyporus lucidus* var. *japonicus* (Fr.) Cleland & Cheel
	=*Ganoderma lucidum* subsp. *lucidum* (Curtis) P. Karst.
	=*Ganoderma lucidum* var. *orbiformis* Steyaert
	=*Ganoderma lucidum* f. *naiae* Chona & Munjal
	=*Grifola lucida* (Curtis) Gray
	=*Ganoderma lucidum* var. *lucidum* (Curtis) P. Karst.
	=*Fomes lucidus* f. *lucidus* (Curtis) Sacc.
	=*Fomes lucidus* var. *lucidus* (Curtis) Sacc.
	=*Polyporus lucidus* f. *lucidus* (Curtis) Fr.
	=*Boletus castaneus* f. *castaneus* Weber
	=*Ganoderma lucidum* f. *lucidum* (Curtis) P. Karst.
	=*Polyporus lucidus* var. *lucidus* (Curtis) Fr.
	=*Polyporus lucidus* (Curtis) Fr.
	=*Ganoderma nitens* Lázaro Ibiza
	=*Agarico-igniarium trulla* Paulet
	=*Agaricus lignosus* Lam.
	=*Boletus crustatus* J.J. Planer
	=*Boletus laccatus* Timm
	=*Boletus verniceus* Brot.

(continued)

Table 2.2 (continued)

Current names	Synonyms
	=*Boletus vernicosus* Bergeret
	=*Ganoderma lucidum* var. *badium* Pat.
	=*Ganoderma ostreatum* var. *hemicycla* Lázaro Ibiza
	=*Ganoderma pseudoboletus* (Jacq.) Murrill
	=*Ganoderma resinaceum* var. *martellii* Bres.
	=*Phaeoporus lucidus* (Curtis) J. Schröt.
	=*Polyporus lucidus* var. *nicotianae* Inglese
	=*Scindalma japonicum* (Fr.) Kuntze
	=*Placodes lucidus* (Curtis) Quél.
	=*Boletus rugosus* Jacq.
	=*Boletus castaneus* Weber
	=*Ganoderma ostreatum* var. *ostreatum* Lázaro Ibiza
	=*Ganoderma lucidum* f. *boninense* Pat.
	=*Ganoderma lucidum* f. *alneum* Bourdot & Galzin
	=*Ganoderma lucidum* f. *martellii* (Bres.) Bourdot & Galzin
	=*Polyporus laccatus* subsp. *semipatera* Pers.
	=*Boletus ramulosum* var. *flabelliformis* (Leyss.) J.F. Gmel.
	=*Boletus supinus* var. *castaneus* (Weber) J.F. Gmel.
	=*Polyporus laccatus* subsp. *laccatus* (Timm) Pers.
	=*Ganoderma lucidum* f. *noukahivense* Pat.
	=*Ganoderma laccatum* (Timm) Pat.
Ganoderma lusambilaense Steyaert	
Ganoderma luteicinctum Corner	
Ganoderma luteomarginatum J.D. Zhao, L.W. Hsu & X.Q. Zhang	
Ganoderma luteum Steyaert	
Ganoderma magniporum J.D. Zhao & X.Q. Zhang	
Ganoderma maitlandii Steyaert	=*Ganoderma maitlandii* var. *ellipsosporum* Steyaert
	=*Ganoderma maitlandii* var. *maitlandii* Steyaert
Ganoderma malayanum Steyaert	
Ganoderma manoutchehrii Steyaert	
Ganoderma martinicense Welti & Courtec.	
Ganoderma mbrekobenum E.C. Otto, Blanchette, Held, C.W. Barnes & Obodai	

(continued)

Table 2.2 (continued)

Current names	Synonyms
Ganoderma mediosinense J.D. Zhao	
Ganoderma megalosporum Steyaert	
Ganoderma meijiangense J.D. Zhao	
Ganoderma melanophaeum Steyaert	
Ganoderma meredithiae Adask. & Gilb.	
Ganoderma microsporum R.S. Hseu	
Ganoderma miniatocinctum Steyaert	
Ganoderma mirabile C.J. Humphrey	=*Ganoderma bakeri* Pat.
	=*Fomes fuscopallens* Bres.
	=*Fomes mirabilis* Lloyd
Ganoderma mirivelutinum J.D. Zhao	
Ganoderma mizoramense Zothanz., Blanchette, Held & C.W. Barnes	
Ganoderma multicornum Ryvarden	
Ganoderma multipileum Ding Hou	
Ganoderma multiplicatum (Mont.) Pat.	=*Ganoderma subamboinense* var. *laevisporum* Bazzalo & J.E. Wright
	=*Fomes multiplicatus* (Mont.) Sacc.
	=*Ganoderma subamboinense* Henn.
	=*Ganoderma multiplicatum* var. *vitalii* Steyaert
	=*Ganoderma subamboinense* var. *subamboinense* Henn.
	=*Ganoderma multiplicatum* var. *multiplicatum* (Mont.) Pat.
	=*Scindalma multiplicatum* (Mont.) Kuntze
	=*Polyporus multiplicatus* Mont.
Ganoderma mutabile Y. Cao & H.S. Yuan	
Ganoderma namutambalaense Steyaert	
Ganoderma neogibbosum Welti & Courtec.	
Ganoderma neojaponicum Imazeki	
Ganoderma nigrolucidum (Lloyd) D.A. Reid	=*Polyporus nigrolucidus* Lloyd
Ganoderma nitidum Murrill	=*Polyporus nitidus* (Murrill) Overh.
Ganoderma ochrolaccatum (Mont.) Pat.	=*Fomes ochrolaccatus* var. *cornucopiae* (Henn.) Sacc.
	=*Ganoderma buissonii* Pat.
	=*Fomes ochrolaccatus* (Mont.) Pat.
	=*Fomes ochrolaccatus* var. *ochrolaccatus* (Mont.) Pat.
	=*Ganoderma ochrolaccatum* var. *cornucopiae* Henn.
	=*Scindalma ochrolaccatum* (Mont.) Kuntze
	=*Polyporus ochrolaccatus* Mont.

(continued)

Table 2.2 (continued)

Current names	Synonyms
	=*Ganoderma ochrolaccatum* var. *ochrolaccatum* (Mont.) Pat.
Ganoderma oerstedii (Fr.) Torrend	=*Fomes oerstedii* (Fr.) Cooke
	=*Scindalma oerstedii* (Fr.) Kuntze
	=*Polyporus oerstedii* Fr.
Ganoderma orbiforme (Fr.) Ryvarden	=*Ganoderma boninense* Pat.
	=*Ganoderma mastoporum* (Lév.) Pat.
	=*Ganoderma limushanense* J.D. Zhao & X.Q. Zhang
	=*Ganoderma densizonatum* J.D. Zhao & X.Q. Zhang
	=*Fomes lucidus* f. *noukahivensis* Sacc.
	=*Fomes mastoporus* (Lév.) Cooke
	=*Fomes lucidus* f. *boninensis* Sacc.
	=*Fomes orbiformis* (Fr.) Cooke
	=*Ganoderma subtornatum* Murrill
	=*Ganoderma lucidum* var. *orbiformis* (Fr.) Rick
	=*Fomes cupreus* Cooke
	=*Ganoderma tornatum* var. *subtornatum* (Murrill) P.W. Graff
	=*Elfvingia mastopora* (Lév.) Imazeki
	=*Fomes mastoporus* f. *rugosus* G. Cunn.
	=*Ganoderma noukahivense* Pat.
	=*Fomes mastoporus* f. *mastoporus* (Lév.) Cooke
	=*Polyporus orbiformis* Fr.
	=*Fomes subtornatus* (Murrill) Lloyd
	=*Scindalma cupreum* (Cooke) Kuntze
	=*Scindalma mastoporum* (Lév.) Kuntze
	=*Scindalma orbiforme* (Fr.) Kuntze
	=*Polyporus mastoporus* Lév.
	=*Polyporus cupreus* Fr.
	=*Ganoderma cupreum* (Cooke) Bres.
	=*Ganoderma lucidum*
Ganoderma oregonense Murrill	=*Ganoderma nevadense* Murrill
	=*Fomes oregonensis* (Murrill) Sacc. & Traverso
	=*Ganoderma sequoiae* Murrill
Ganoderma oroleucum Pat. & Har.	
Ganoderma ostracodes Pat.	
Ganoderma parvigibbosum Welti & Courtec.	

(continued)

Table 2.2 (continued)

Current names	Synonyms
Ganoderma parviungulatum J.D. Zhao & X.Q. Zhang	
Ganoderma perzonatum Murrill	
Ganoderma petchii (Lloyd) Steyaert	=*Fomes petchii* Lloyd
Ganoderma pfeifferi Bres.	=*Ganoderma pfeifferi* var. *borneense* Corner
	=*Ganoderma applanatum* var. *laccatum* (Sacc.) Rea
	=*Ganoderma applanatum* f. *laccatum* (Sacc.) Golovin
	=*Polyporus laccatus* Kalchbr.
	=*Fomes laccatus* Sacc.
	=*Fomes pfeifferi* (Bres.) Sacc.
	=*Fomes cupreolaccatus* Kalchbr. ex Ferd. & C.A. Jørg.
	=*Ganoderma laccatum* (Sacc.) Bourdot & Galzin
	=*Ganoderma cupreolaccatum* Kalchbr. ex Z. Igmándy
	=*Ganoderma pfeifferi* var. *pfeifferi* Bres.
	=*Scindalma laccatum* (Sacc.) Kuntze
	=*Scindalma pfeifferi* (Bres.) Kuntze
	=*Polyporus cupreolaccatus* Kalchbr.
Ganoderma philippii (Bres. & Henn. ex Sacc.) Bres.	=*Fomes pseudoferreus* Wakef.
	=*Fomes philippii* Bres. & Henn. ex Sacc.
	=*Ganoderma pseudoferreum* (Wakef.) Overeem & B.A. Steinm.
	=*Scindalma philippii* (Bres. & Henn. ex Sacc.) Kuntze
Ganoderma piceum (Ces.) Ryvarden	=*Polyporus piceus* Ces.
	=*Fomes piceus* (Ces.) Sacc.
	=*Scindalma piceum* (Ces.) Kuntze
Ganoderma platense Speg.	
Ganoderma plicatum Pat.	
Ganoderma podocarpense J.A. Flores, C.W. Barnes & Ordoñez	
Ganoderma polychromum (Copel.) Murrill	=*Polyporus polychromus* Copel.
Ganoderma polymorphum Cleland	
Ganoderma puglisii Steyaert	
Ganoderma pulchella Bres.	
Ganoderma pygmoideum Steyaert	
Ganoderma ramosissimum J.D. Zhao	
Ganoderma ravenelii Steyaert	

(continued)

Table 2.2 (continued)

Current names	Synonyms
Ganoderma renii S.C. He	
Ganoderma resinaceum Boud.	=*Ganoderma praelongum* Murrill
	=*Fomes subperforatus* (G.F. Atk.) Sacc. & Traverso
	=*Fomes chaffangeonii* (Pat.) Sacc.
	=*Fomes resinaceus* (Boud.) Sacc.
	=*Ganoderma sessile* Murrill
	=*Fomes sessilis* (Murrill) Sacc. & D. Sacc.
	=*Ganoderma chaffangeonii* Pat.
	=*Ganoderma pulverulentum* Murrill
	=*Ganoderma subperforatum* G.F. Atk.
	=*Ganoderma subincrustatum* Murrill
	=*Ganoderma argillaceum* Murrill
	=*Ganoderma subfornicatum* Murrill
	=*Polyporus argillaceus* (Murrill) Overh.
	=*Polyporus pulverulentus* (Murrill) Overh.
	=*Ganoderma resinaceum* var. *alneum* Bourdot & Galzin
	=*Polyporus subincrustatus* (Murrill) Seaver & Chardón
	=*Ganoderma subtuberculosum* Murrill
	=*Ganoderma resinaceum* f. *quercinum* Dzhaf.
	=*Fomes resinaceus* var. *resinaceus* (Boud.) Sacc.
	=*Ganoderma resinaceum* var. *resinaceum* Boud.
	=*Ganoderma resinaceum* f. *resinaceum* Boud.
	=*Ganoderma areolatum* Murrill
	=*Ganoderma perturbatum* (Lloyd) Torrend
	=*Fomes areolatus* (Murrill) Murrill
	=*Friesia resinacea* (Boud.) Lázaro Ibiza
	=*Polyporus perturbatus* Lloyd
	=*Polyporus sessilis* (Murrill) Lloyd
	=*Scindalma chaffangeonii* (Pat.) Kuntze
	=*Scindalma resinaceum* (Boud.) Kuntze
	=*Mensularia vernicosa* var. *vernicosa* Lázaro Ibiza
	=*Ganoderma lucidum* subsp. *resinaceum* (Boud.) Bourdot & Galzin
Ganoderma reticulatosporum (Van der Byl) D.A. Reid	=*Polyporus reticulatosporus* Van der Byl

(continued)

Table 2.2 (continued)

Current names	Synonyms
Ganoderma rhacodes Pat.	
Ganoderma rothwellii Steyaert	
Ganoderma rotundatum J.D. Zhao, L.W. Hsu & X.Q. Zhang	
Ganoderma rufoalbum (Bres. & Pat.) Pat.	=*Ceriomyces rufoalbus* (Bres. & Pat.) Sacc.
	=*Ptychogaster rufoalbus* Bres. & Pat.
Ganoderma ryvardenii Tonjock & Mih	
Ganoderma sanmingense J.D. Zhao & X.Q. Zhang	
Ganoderma sarasinii Steyaert	
Ganoderma sculpturatum (Lloyd) Ryvarden	=*Fomes sculpturatus* Lloyd
Ganoderma septatum Steyaert	
Ganoderma sessiliforme Murrill	=*Fomes sessiliformis* (Murrill) Murrill
Ganoderma shandongense J.D. Zhao & L.W. Xu	
Ganoderma shangsiense J.D. Zhao	
Ganoderma sichuanense J.D. Zhao & X.Q. Zhang	=*Ganoderma lingzhi* Sheng H. Wu, Y. Cao & Y.C. Dai
Ganoderma silveirae Torrend	
Ganoderma simaoense J.D. Zhao	
Ganoderma simulans Wakef.	
Ganoderma sinense J.D. Zhao, L.W. Hsu & X.Q. Zhang	
Ganoderma soniense Steyaert	
Ganoderma soyeri Steyaert	
Ganoderma staneri Steyaert	
Ganoderma steyaertianum B.J. Sm. & Sivasith.	
Ganoderma stipitatum (Murrill) Murrill	=*Fomes parvulus* (Murrill) Sacc. & D. Sacc.
	=*Fomes stipitatus* Murrill
	=*Ganoderma parvulum* Murrill
	=*Ganoderma bibadiostriatum* Steyaert
Ganoderma stratoideum S.C. He	
Ganoderma sublucidum (Beeli) Steyaert	=*Polyporus lucidus* f. *sublucidus* (Beeli) E.K. Cash
	=*Polyporus sublucidus* Beeli
Ganoderma subrenatum (Murrill) Sacc. & Trotter	=*Amauroderma subrenatum* Murrill
	=*Polyporus subrenatus* (Murrill) Lloyd
Ganoderma subresinosum (Murrill) C.J. Humphrey	=*Amauroderma subresinosum* (Murrill) Corner
	=*Polyporus mamelliporus* Beeli
	=*Fomes subresinosus* Murrill
	=*Trachyderma subresinosum* (Murrill) Imazeki

(continued)

Table 2.2 (continued)

Current names	Synonyms
	=*Magoderna subresinosum* (Murrill) Steyaert
Ganoderma substipitata Bres.	
Ganoderma subumbraculum Imazeki	
Ganoderma tenue J.D. Zhao, L.W. Hsu & X.Q. Zhang	
Ganoderma testaceum (Cooke) Pat.	=*Fomes testaceus* Cooke
	=*Scindalma testaceum* (Cooke) Kuntze
	=*Polyporus testaceus* Lév.
Ganoderma theaecola J.D. Zhao	
Ganoderma tibetanum J.D. Zhao & X.Q. Zhang	
Ganoderma torosum Steyaert	
Ganoderma trengganuense Corner	
Ganoderma triangulum J.D. Zhao & L.W. Hsu	
Ganoderma tropicum (Jungh.) Bres.	=*Fomes tropicus* (Jungh.) Cooke
	=*Polyporus tropicus* Jungh.
	=*Scindalma tropicum* (Jungh.) Kuntze
Ganoderma trulla Steyaert	
Ganoderma trulliforme Steyaert	=*Ganoderma trulliforme* var. *radicicola* Corner
	=*Ganoderma trulliforme* var. *trulliforme* Steyaert
Ganoderma tsugae Murrill	=*Fomes tsugae* (Murrill) Sacc. & D. Sacc.
	=*Polyporus tsugae* (Murrill) Overh.
	=*Ganoderma tsugae* var. *jannieae* Wasser, Zmitr. & M. Didukh
	=*Ganoderma tsugae* var. *tsugae* Murrill
Ganoderma tuberculosum Murrill	
Ganoderma turbinatum Ipulet & Ryvarden	
Ganoderma umbrinum Bres.	
Ganoderma ungulatum J.D. Zhao & X.Q. Zhang	
Ganoderma valesiacum Boud.	=*Fomes valesiacus* (Boud.) Sacc. & P. Syd.
	=*Ganoderma lucidum* var. *valesiacum* (Boud.) Lloyd
	=*Polyporus valesiacus* (Boud.) Lloyd
	=*Scindalma valesiacum* (Boud.) Kuntze
	=*Ganoderma lucidum* subsp. *valesiacum* (Boud.) Bourdot & Galzin
Ganoderma vanheurnii Steyaert	
Ganoderma vanmeelii Steyaert	
Ganoderma vivianimercedianum M. Torres	
Ganoderma weberianum (Bres. & Henn. ex Sacc.) Steyaert	=*Ganoderma rivulosum* Pat. & Har.

(continued)

Table 2.2 (continued)

Current names	Synonyms
	=*Polyporus weberianus* (Bres. & Henn. ex Sacc.) Trotter
	=*Polyporus subcapucinus* Bres.
	=*Fomes weberianus* Bres. & Henn. ex Sacc.
	=*Coltricia weberiana* (Bres. & Henn. ex Sacc.) G. Cunn.
	=*Phylloporia weberiana* (Bres. & Henn. ex Sacc.) Ryvarden
	=*Scindalma weberianum* (Bres. & Henn. ex Sacc.) Kuntze
Ganoderma wiiroense E.C. Otto, Blanchette, C.W. Barnes & Held	
Ganoderma williamsianum Murrill	=*Elfvingia williamsiana* (Murrill) Imazeki
Ganoderma wuhuense X.F. Ren	
Ganoderma wuzhishanense T.C. Wen, Hapuar. & K.D. Hyde	
Ganoderma wynaadense Steyaert	
Ganoderma xanthocreas Pat.	
Ganoderma xingyiense S.C. He	
Ganoderma xylonoides Steyaert	
Ganoderma zhenningense S.C. He	
Ganoderma zonatum Murrill	=*Ganoderma sulcatum* Murrill
	=*Ganoderma tumidum* Bres.
	=*Fomes zonatus* (Murrill) Sacc. & D. Sacc.
	=*Polyporus lucidus* var. *zonatus* (Murrill) Overh.

With the development of molecular phylogenies in the late twentieth century, species concept hypotheses were tested to determine the relatedness amongst the nuanced morphological variabilities of the laccate *Ganoderma* taxa. In 1995, Moncalvo et al. constructed a phylogeny of the rDNA, which was the universally accepted locus at that time, and found five major clades of the laccate species amongst the 29 isolates tested [75]. It turned out that *G. lucidum* was not a monophyletic species, and further work needed to be done to clarify this taxonomic problem. They also found that *G. resinaceum* from Europe and the North American "*G. lucidum*", which Adaskaveg and Gilbertson found to be biologically compatible in vitro [76], did not cluster together [75]. Moncalvo et al. rejected biological species complexes as a sole tool to distinguish a taxon and suggested using a combination between biological and phylogenetic species concepts to define unique *Ganoderma* taxa [75].

Phylogenetic analysis using DNA sequence information has helped to clarify our understanding of the relationships amongst *Ganoderma* species [77, 78]. The genus was divided into six monophyletic groups – *G. colossus*, *G. applanatum*, *G. tsugae*,

Asian *G. lucidum*, *G. meredithiae* Adask. & Gilb. and *G. resinaceum* – in the study of Hong & Jung [55].

Nowadays, there are four accepted subgenera under *Ganoderma* in the Index Fungorum database: *Ganoderma*, *Trachyderma*, *Humphrey* and *Haddowia*.

2.2.4 Phylogenetic Study of Ganoderma

Moncalvo et al. and Bae et al. firstly used the internal transcribed spacer (ITS) and large subunit (LSU) rDNA sequences to distinguish species of *Ganoderma* [74, 79, 80]. Later many DNA regions have been adopted into the genus *Ganoderma*, such as nrDNA-LSU [81, 82], nrDNA-SSU [83], nrDNA-ITS [80, 84, 85], mitDNA-SSU [55, 56, 86], β-tubulin gene [87], RNA polymerase II subunit and intergenic spacers [66], RNA polymerase I subunit and translation elongation factor 1-α [67].

A variety of molecular biotechnologies have been used to study the genetic diversity in *Ganoderma*, for example isozyme analysis [88], AFLP (amplified fragment length polymorphism) [89, 90], ISSR (inter simple sequence repeat) [91], RAPD (random amplified polymorphic DNA) [91, 92], RFLP (restricted fragment length polymorphisms) [90], SCAR (sequence characterised amplified region) [93, 94], SNP (single-nucleotide polymorphism) [93], SRAP (sequence-related amplified polymorphism) [95] and SSCP (single-strand conformational polymorphism) [96].

These different molecular markers had been applied to different taxonomic classifications of *Ganoderma*, which made great progress and provided a lot of information for the further study of *Ganoderma*.

2.2.4.1 Genera and Subgenera in Ganodermataceae

Moncalvo et al. indicated that *Amauroderma* was a sister clade of *Ganoderma*, and the subgenus *Elfvingia* was monophyletic within *Ganoderma* based on the phylogenetic analysis from data 25S ribosomal RNA gene and ITS sequence [79]. In addition, the low sequence variation observed in the 25S ribosomal RNA gene within *Ganoderma* species suggested that *Ganoderma* was a newly originated genus [79]. Gottlieb et al. clarified subgen. *Elfvingia* and subgen. *Ganoderma* were monophyletic based on morphology and molecular methods, which included the use of PCR coupled to restriction enzyme digestions, single-strand conformational polymorphism (SSCP) and direct sequencing to assay rDNA polymorphism [95]. Smith and Sivasithamparam supported the retention of *Elfvingia* and *Ganoderma* as subgenera of *Ganoderma* by ITS sequences from 23 strains of five *Ganoderma* species and 38 sequences from EMBL database [97]. Hong and Jung proved that *Tomophagus* is a taxonomically appropriate genus, and subgenus *Elfvingia* was a monophyletic group, based on the phylogenetic analysis of mtSSU rDNA sequences [55]. *Amauroderma, Ganoderma* and *Perenniporia* Murrill were placed as closely related groups with a moderate support, based on the combined RPB2-LSU analysis [98]. Zheng et al. investigated the closely related *Ganoderma* strains and clarified genetic relations at molecular level, using 14 primer combinations of amplified fragment length polymorphism (AFLP) and internal transcribed spacer (ITS) PCR-RFLP. The cluster analysis revealed that the taxonomical system of subgenus *Ganoderma* is composed of sect. *Ganoderma* and sect. *Phaeonema* [90]. Richer et al. provided an overview of *Ganoderma* based on the ITS sequences, and the subgenus *Elfvingia* is clearly separated from the other *Ganoderma* species, of which the Asian and European groups of *G. lucidum* appear in distant clades [84].

2.2.4.2 Species Relationship in *Ganoderma*

Moncalvo et al. summarised the genus *G. lucidum* complex based on ITS region and divergent domain D2 of the LSU gene, revealing that collections named *G. lucidum* in North America and Asia are not conspecific with European *G. lucidum* [75]. The sister group of European *G. lucidum* is an Argentinian taxon labelled *G. oerstedii* (Fr.) Torrend. North American *G. lucidum* is related to a Formosan isolate identified as *G. boninense*. *Ganoderma tsugae* is absent from Taiwan of China and probably from Japan and mainland of China [75]. Smith and Sivasithamparam studied the ITS region of *Ganoderma* species from Australia, which showed *G. adspersum* (Schulzer) Donk, a European species, was not synonymous with *G. australe* (Fr.) Pat. from Australia; *Ganoderma incrassatum* (Berk.) Bres. from northern Australia formed a clade with isolates named *G. gibbosum* and *G. austral* from China; *Ganoderma microsporum* R.S. Hseu was a synonym of *G. weberianum* [97]. Hong et al. studied the secondary structure of the SSU domains [99]. The results showed that the structure of variable domains can be used as a valuable marker and that the SSU region has three times more information than ITS among *Ganoderma* species. Their studies also revealed that the information of the SSU region is concentrated mostly in five variable domains (V1, V4, V5, V6 and V9). And the presence of chlamydospores is an important characteristic for species identification, as it is also corroborated by molecular phylogenetic studies [99]. Only a few *Ganoderma* species, such as *G. colossus, G. subamboinense* and *G. weberianum,* contain chlamydospores in their basidiomata [55, 56].

Moncalvo and Buchanan stated that ITS sequence can distinguish the single branch of geographic evolution of the *G. applanatum–austral* complex [85]. Glen et al. investigated *Amauroderma* and *Ganoderma* species associated with root-rot disease of *Acacia mangium* plantation trees in Southeast Asia using ITS sequence, including *G. philippii* (Bres. & Henn. ex Sacc.) Bres., *G. mastoporum, G.* aff. *Steyaertanum, G. australe* and *A. rugosum.* The phylogenetic results showed that *G. mastoporum* (Lév.) Pat. from Southeast Asia and *G. cupreum* (Cooke) Bres. from Australia clustered into the same evolutionary branch [100]. In the same year, *G. lucidum* was confirmed as a name mistakenly applied to oriental collections, and *G. multipileum* Ding Hou was found in the study as the earliest valid name for "*G. lucidum*" known from tropical Asia [101]. Wang et al. identified the widely cultivated *Ganoderma* in China based on three-gene combined analyses (ITS+IGS + RPB2) [66]. Cao et al. added gene fragments of RPB1 and TEF1-α to clarify the species of *Ganoderma* [67]. Zhang et al. revealed the intraspecific variation and phylogenetic relationships of *G. lucidum* using ITS1 secondary structure analysis and single-nucleotide polymorphism [102].

2.3 Biological Characteristics of *Ganoderma* Species

2.3.1 Morphological Variations

2.3.1.1 Macromorphology

Plieus: The plieus consists of context, cutis and tube, presenting different colours, such as yellowish-white, light brown, yellowish-brown, brown, reddish-brown, dark brown, puce and black, and different shapes, for example flabelliform, suborbicular, cochlear, reniform, hippocrepiform and other irregular shapes. There are also differences in the size of the pileus. In addition, others such as laccate or nonlaccate, with radial furrows or with concentrically sulcate zones, are also important criteria of identification.

Context: The context is the subcutaneous part of the cutis. The colour of the context varies, for example pale white, white, yellowish-brown, dark brown, cinnamon. Context usually has one layer or several layers. Some species have one or more chitin lines in context, producing gasterospore sometimes in the context.

Pore surface: The pore surface is initially white or cream-coloured, becoming yellowish-brown, dark brown or puce when mature.

Tube: The tube is composed of hyphae, producing basidiospores. The colour of the tube is yellowish-brown, brown, dark brown or puce. Tubes exist layer by layer; the length and the thickness of the tube are also important characteristics.

Stipe: The stipe is composed of hyphae and generally divided into sessile and substipitate to stipitate based on the presence or absence of the stipe. Attachment type of the stipe on pileus are varied from lateral to nearly central. Additional shape, length, diameter, colour and laccate or nonlaccate are often of value in taxonomy.

2.3.1.2 Micromorphology

Hyphal system: The hyphal system includes monomitic, dimic and trimitic. Most species in the genus *Ganoderma* are trimitic, including generative hyphae, skeletal hyphae and binding hyphae. Generally, generative hyphae are hyaline, thin-walled, with clamp connections. Skeletal hyphae are darker than generative hyphae, yellowish-brown, thick-walled to solid. Binding hyphae are hyaline, branched, thick-walled to solid.

Basidia: The basidia are near hyaline, thin-walled, short and strong, spherical to clavate, producing four sterigmata each basidium.

Basidiospores: The basidiospores are double-walled; walls are externally colourless, internally dark brown. Basidiospores are ovoid to oblong, truncate or not at the apex. The length-width ratio of double walls and the surface structures are all the stable characters.

Cutis: The cutis has types of amauroderm, hymenioderm, plecoderm, trichoderm and characoderm. The shape, colour, length and weather thickness at the top of the cutis have obvious morphological differences in identification of species.

Gasterospores: The gasterospores are spherical or fusiform, thick-walled in asexual morph, which exist in the context and tube [65, 103–106].

2.3.2 Ecology and Distribution

2.3.2.1 Distribution

The distribution of *Ganoderma* species covers all over the world, especially in tropical, subtropical and temperate areas [104]. Amongst the totally 213 species in genus *Ganoderma*, 100 species were reported from China according to the database of Checklist of Fungi in China (http://www.fungalinfo.net, accessed January, 2019). *Ganoderma* is also mostly distributed in tropical and subtropical areas in China. At the provincial level, Hainan, Hong Kong, Tibet and Yunnan have the greatest biodiversity of more than 50 species. Anhui, Gansu and Zhejiang have the least of only four species.

2.3.2.2 Habitat

Ganoderma species have a worldwide distribution in both tropical and temperate geographical regions, growing as a parasite or saprotroph on a wide variety of trees. In the wild, *Ganoderma* is often solitary or gregarious at the dead trunks, fallen wood, rotten wood, rotten roots and stumps of deciduous, coniferous or mingled forest. Most of the species grow on deciduous trees such as *Alnus* sp., *Betula* sp., *Castanea* sp., *Castanopsis* sp., *Cyclobalanopsis* sp., *Populus* sp., *Salix* sp., *Quercus* sp. and *Ulmus* sp. Some species grow on specific habitat, for example, *G. achalceum* mainly parasitises on *Dacrycarpus* sp., *G. mirabile* can grow at the higher position of the trunk, *G. philippi* mainly grows on *Heave* sp., *G. tropicum* mostly parasitises on *Acacia* sp., *G. tsugae* only grows on coniferous trees, and *G. trulla* grows on *Castanopsis hystrix* of Hainan island and Indonesian islands. In addition, *G. applanatum*, *G. australe*, *G. lobatum*, *G. tropicum*, *G. tsugae*, *G. tsunodae* and *G. weberianum* can parasitise not only on rotten trees, but also living trees.

2.3.3 Life History

Sharing the similar life history as Basidiomycota, *Ganoderma* species tend to have mutually indistinguishable, compatible haploids which are usually mycelia being composed of filamentous hyphae. Typically, haploid mycelia fuse via plasmogamy and then the compatible nuclei migrate into each other's mycelia and pair up with the resident nuclei. Karyogamy is delayed, so that the compatible nuclei remain in pairs, called a dikaryon. The hyphae are then said to be dikaryotic. Conversely, the haploid mycelia are called monokaryons. The dikaryons can be long-lived, lasting years, decades or centuries. The monokaryons are neither male nor female. This results in the fact that following meiosis, the resulting haploid basidiospores and resultant monokaryons have nuclei that are compatible with their sister basidiospores (and their resultant monokaryons) because the mating genes must differ for them to be compatible. The sexual system of some *Ganoderma* species has been shown to be tetrapolar (i.e. these species have different mating type alleles at two unlinked loci) and heterothallic [85, 106, 107].

The maintenance of the dikaryotic status in dikaryons is facilitated by the formation of clamp connections that physically appear to help coordinate and re-establish pairs of compatible nuclei following synchronous mitotic nuclear divisions. In a life cycle the long-lasting dikaryons periodically (seasonally or occasionally) produce basidia, the specialised usually club-shaped end cells, in which a pair of compatible nuclei fuse (karyogamy) to form a diploid cell. Meiosis follows shortly with the production of four haploid nuclei that migrate into four external, usually apical basidiospores. Typically the basidiospores are ballistic; hence, they are sometimes also called ballistospores. In most species, the basidiospores disperse and each can start a new haploid mycelium, continuing the life cycle. Basidia are microscopic but they are often produced on or in multicelled large fructifications called basidiocarps or basidiomes, or fruit bodies. Ballistic basidiospores are formed on sterigmata which are tapered spine-like projections on basidia and are typically curved, like the horns of a bull.

In summary, meiosis takes place in a diploid basidium. Each one of the four haploid nuclei migrate into its own basidiospore. The basidiospores are ballistically discharged and start new haploid mycelia called monokaryons. There are no males or females; rather there are compatible thalli with multiple compatibility factors. Plasmogamy between compatible individuals leads to delayed karyogamy leading to establishment of a dikaryon. The dikaryon is long lasting but ultimately gives rise to either fruit bodies with basidia or directly to basidia without fruit bodies. The paired dikaryon in the basidium fuse (i.e. karyogamy takes place). The diploid basidium begins the cycle again.

2.3.4 *Conditions and Requirements for Growth*

2.3.4.1 Nutritional Condition

Ganoderma is a cosmopolitan genus of white rot fungi of economic importance, and some of the species are recognised as a potentially important source of lignin-degrading enzymes which have been shown to selectively delignify wood. *Ganoderma* can grow on rotten trees in broad-leaved and coniferous forests. In wood-log cultivation, the necessary nutrients are obtained from the phloem and xylem. In substitute cultivation, the carbon source mainly comes from the sawdust of broad-leaved tree, cottonseed husk, corn cob and soybean straw, and the nitrogen source from bran. The more reasonable ratio of main and auxiliary materials is of great importance for the cultivation of *Ganoderma* with high yield and great quality [70].

2.3.4.2 Environmental Condition

Temperature

The hyphae of *Ganoderma* species can grow between 5 and 35 °C. Hyphae can withstand low temperature of 0 °C, maintaining the lowest physiological activity, but will stop growing and will return to normal growth when the temperature rises to an appropriate level [104]. The optimum temperature for the growth of hyphae in the matrix is 25–30 °C, and 24–26 °C is the suitable temperature for the germination of basidiospore. Fruit body can differentiate between 24 and 28 °C and develop normally at 25–30 °C. Intense temperature variations can cause pileus deformity [70].

Moisture

Ganoderma species is a hygrophilous fungus. Relative humidity in air should be maintained between 65% and 70% during the hyphae growth, which will prevent the moisture in the matrix from evaporating. The relative humidity of air should be maintained between 85% and 90% during the development and growth stages of fruit bodies. If the relative humidity is less than 60%, serious evaporation of water in the matrix and hyphae would cause water shortage and growth inhibition of fruit body. If the relative humidity is lower than 45%, the growth of the hyphae would stop and the young fruit bodies die. Besides, if the relative humidity is higher than 95%, the respiration of hyphae and fruit body would be blocked because of the decreasing oxygen content in the air, resulting in hyphae autolysis and the fruit body decaying to death.

The water content of the matrix should be 37%–40% in the wood-log cultivation. In substitute cultivation, 60%–65% of the water content is suitable. When the water content in the culture medium is higher than 80%, the low oxygen content easily causes the death of the hyphae.

Oxygen and Carbon Dioxide

Ganoderma is an aerobic fungus. If the oxygen is insufficient, the hyphae grow slowly during the stage of the hypha growth. When hypoxia is serious, it stops growing. Under natural conditions, the concentration of carbon dioxide in air is 0.03%, and the hyphae of *Ganoderma* could grow normally. Increasing the concentration of carbon dioxide can promote the growth of hyphae, and the growth rate can be accelerated by more than two to three times when the concentration of carbon dioxide increases to 0.1%–10% [104]. The development of fruit bodies is sensitive to carbon dioxide in the air. The suitable content of carbon dioxide in the air for the differentiation and growth of fruit bodies is 0.03%–0.1%. When the concentration of carbon dioxide is higher than normal, the shape of the fruit body will be affected, the growth will be inhibited, and the stipe will become antler-like branches and may even not form a fruit body.

Acidity and Alkalinity

Ganoderma species are suitable for acidic growth conditions. Hyphae can grow in the range of pH 3–9, and the most suitable pH value for hypha growth is 4.5–5.2. The growth rate of hyphae will slow down when pH is 8 and will stop growing when pH is higher than 9. Under alkaline conditions, the solubility of inorganic ions such as calcium ion and magnesium ion will increase, which can inhibit the activities of various enzymes, synthesis of vitamins and normal metabolic activities.

Illumination

Ganoderma is a heterotrophic fungus, which cannot carry out photosynthesis. The hyphae can grow normally under dark conditions. Blue and purple light in visible light have obvious inhibitory effect on hypha growth. Ultraviolet radiation at 260–265 nm can destroy the RNA, DNA and nucleoprotein in hyphae, and the hyphae can be killed by ultraviolet radiation about 30 min. Direct sunlight is harmful to hyphae. The stronger the light is, the more harmful to hyphae. The red light at 570–920 nm is harmless to growth. The differentiation of the fruit body requires induction of blue light (about 400–500 nm). In dark or weak light condition (illumination: 20–1000 lx), only stipe grows and no pileus forms. When the illumination intensity reaches above 1500 lx, the bud grows fast and can form a normal pileus. The stipe has phototaxis, and the light in one direction can promote the stipe to grow in the direction of strong light and become too long. The growth of the basidium is photophobic. Purple light and ultraviolet are harmful to its growth particularly, but it needs to be induced by purple light when hyphae differentiate into basidium. Blue and purple light are harmful to the development of basidiospore. During the cultivation process, the light around the mushroom shed should be as uniform as possible, and the cultivation bags and wood logs should not be moved frequently to prevent the pileus from deformity [70, 104].

2.4 Cultivations of *Ganoderma* Species

Artificial cultivation of *Ganoderma* has spread widely in China. Fujian, Guangdong, Guangxi, Guizhou, Yunnan and Zhejiang provinces in the south of China are still the main producing areas of *Ganoderma*, and there are also considerable *Ganoderma* yields in the north, for example Great Khingan, Jilin and Shandong provinces. Tibet, Xinjiang and other provinces in southwest China have become the rising stars of *Ganoderma* cultivation areas [108].

Three species, Ling-zhi, Zi-zhi and *G. tsugae*, are recorded in "the list of fungal species that can be used in health food" authorised by the Chinese government. At present, in addition to these three species, other species, for example *G. duropora*, *G. neo-japonicum*, *G. resinaceum* and *G. sessile*, have also been cultivated in large quantities in different places [109].

2.4.1 Selection and Preparation of Substrates for Cultivation

Based on the current knowledge, the cultivation of *Ganoderma* species is generally summarised in the following steps.

2.4.1.1 Wood-Log Cultivation

There are two modes for the wood-log cultivation of *Ganoderma* in China, natural wood-log cultivation and sterile wood-log cultivation. Due to the advantages of its short growth cycle, high biological efficiency and high economic benefits, sterile wood-log cultivation has a wider application.

The procedures for the sterile wood-log cultivation are as follows: selecting species of tree → timely felling → cutting timber and binding → bagging → sterilisation → inoculation → management of hypha growth → arrangement and burying soil → management of fruiting → fruit body harvest and drying → graded packaging [104].

1. *Selecting the Species of Tree and Timely Felling*

Ganoderma is suitable for cultivating on most broadleaf trees. Trees with thick bast, hard material, less heartwood, developed wood ray, no aromatic sputum and rich duct branch are more suitable.

The nutrition and moisture of *Ganoderma* are from mushroom wood. The felling time is very important to improve the quality of mushroom wood. It is appropriate to fell after winter and before the germination period in the spring. The trees are in a dormant period at this time, during when the trees have stored the most nutritious, the sap is not flowing fast, the moisture is suitable, the bark and the xylem are tightly combined, and the bark is not easy to fall off. It is best to cut down the trees on a sunny day rather than in rainy day. The diameter of the trees should not be less than 6 cm and not more than 22 cm, preferably 8–18 cm. During the process of felling and transporting, the bark should not be damaged as much as possible. After felling, the trees are transported to the vicinity of the sterilisation and inoculation sites and stacked in the shade to prevent sun exposure.

2. *Cutting Timber and Bagging*

Timber should be cut about 1 to 2 days before bagging and sterilisation. The lengths of wood logs are about 15–28 cm. The wood log has a diameter of 18–22 cm, each bagged in one plastic bag. The truncation surface needs to be flat to prevent piercing the film bag and causing infection during bagging. For wood log with a small diameter, several sticks of woods can be tied together with bamboo or nylon rope. The tree crotches and branches can also be cut into 20 cm long and bundled into a cylinder shape with a diameter of about 20–25 cm. This method makes rational use of branches and crotches, which is beneficial to protect forest resources. The interspace should be filled with auxiliary materials such as hardwood sawdust. Newly felled wood logs with high moisture content can be dried for 2–3 days. When a 1 cm small crack appears at the centre of the timber section, it can be bagged. At this time, the moisture content of the wood log is approximately 35%–45%. If the wood is too dry, it needs to be immersed in water for 1–2 h before bagging. Firstly, put a layer of sawdust at the bottom of the polyethylene plastic bag, and then put into the wood log, avoiding the wood piercing the bag. Lastly, clean the ends of the bag, seal it with a ring of rubber or a cotton plug, and tie it with a rope.

The following is a formula of the auxiliary materials: sawdust 78%, wheat bran 15%, corn flour 5% and sugar and gypsum 1%. The moisture content varies based on the different woods.

3. *Sterilisation*

Wood log can be sterilised by autoclaving or atmospheric sterilisation. Autoclaving is maintained at 0.15 MPa for 1.5 to 2 h. Atmospheric sterilisation should be maintained for 10 to 12 h above 100 °C, and the temperature should be stable. After sterilisation, the stove door cannot be opened immediately until the temperature naturally drops to 40–50 °C. The bags should be covered with a clean non-woven fabric or plastic film and placed in a clean, dry greenhouse.

4. *Inoculation*

The whole inoculation process should be carried out in a sterile environment, and the inoculation sites and tools should be sterilised strictly. When the temperature of the bag drops to about 30 °C, it can be inoculated. The age of the inoculation strain could be 30–35 days.

Inoculation Method The spawn bags should be cleaned and disinfected in potassium permanganate solution (or benzalkonium bromide solution). When inoculating, remove the cultivar tampon, wipe the bag with 75% ethanol cotton, peel off the surface of the strain with sterilised tweezers, smash the spawn into the size of the grain or soybean, and unpack the bag of wood log. Pick up the spawn by a spoon and pour into the bag, making sure that each piece of wood is exposed to the spawn and then pressurised slightly. The inoculation amount of each wood log is about 5–10 g, and it is still necessary to tighten the ends of the bag after inoculation. If water accumulates in the fungi bag, it should be dumped immediately. If the bag is damaged, it should be put on a spare bag or taped.

5. *Cultivation Sites*

The average temperature in the cultivation sites is suitable in the range of 27–29 °C in July, with annual sunshine amount for about 1700–2000 h. The annual precipitation is 1250–1900 mm. In the south of China, it is advisable to choose an open valley with an altitude of 300–700 m (the highest temperature in summer is below 36 °C, and the average temperature is around 24 °C). Avoid areas where mountain torrents may flow, and it is better to face south or southeast. It is best to have trees all around, which can reduce heat radiation in the summer and increase the oxygen content and humidity of the air around the cultivation site. The reservoir area can also be selected.

In addition, the sites should be chosen away from the village, well drained, open terrain, loose soil, acidic and with convenient water and electricity. The heavy metal content and pesticide residues in the soil and water should be lower than the national sanitary standards. The mushroom field generally can be used for 1–2 years.

One month before burying soil, choose a sunny day to plow the bed, remove weeds and stones, disinfect the soil, and build the ridges after a few days of sun exposure. The height of the ridge is 30–40 cm, the width is 1.5–1.8 m, and the length is determined according to the topography. The drainage channel is built around the mushroom field. The plastic sheds of 1.6–1.8 m high are built on every two ridges and covered with agricultural film, which has the function of shading the sun, adjusting the temperature, concentrating the humidity and carbon dioxide. The shed can be covered with a black shade net. The shade net helps to adjust the temperature and humidity inside the shed, which can be reused for several years and easy to manage.

Some points that need attention: (1) According to the diameter of the wood logs and the growth of the hyphae, the bags are placed separately to ensure that the growth is orderly and easy to manage and harvest. (2) Select a sunny day (temperature above 20 °C) to cover the soil, the site needs to be cleaned in advance, and pay attention to the prevention and control of termites. (3) According to different woods, different spacing is used, the bed surface should be fully utilised, and the fruit bodies of two adjacent woods should not be in contact with each other. The general spacing is 5 cm and the row spacing is 10 cm. (4) After the wood log is discharged neatly, cover the soil about 2 cm and make sure the woods are not exposed. The thickness of the covering soil should be treated as appropriate according to the humidity of the cutting site. It can cover 3–4 cm sandy loam. Then, cover with 5–10 cm dry straw or reed. After the soil is covered, the heavy water is sprayed to fill the interspace between the wood-log bags. The humidity of the soil is: "soil can be held as a clump in the hand and can be sprinkled".

2.4.1.2 Substitute Cultivation

The method of the cultivation of *Ganoderma* species by using plant organic matter instead of wood log is suitable for areas of lacking forest resources in China. Before the 1980s, the substitute cultivation of *Ganoderma* was mainly performed in bottles. After the 1980s, it was changed from bottle to plastic bag. Nowadays, polypropylene or high-pressure polyethylene bags are widely used, which has the advantages of low production cost, high output, good quality, convenient management and more suitable for commercial production.

The procedures of substitute cultivation are the following:

Material selection → burdening → bagging → sterilisation → inoculation → management of hypha growth → management of fruiting → fruit body harvest and drying → graded packaging [104].

1. *Cultivation Season*

The substitute cultivation is generally inoculated in spring, harvested in summer and autumn. The temperature for hyphae and fruit bodies is basically same (25–28 °C) in the northern region. So the bags are usually inoculated at April to May, the fruit bodies grow in June to September, and the cultivation ends in early October. The inoculation time can be advanced to March or April in the southern region.

2. *Substrate Preparation*

The raw materials of culture medium for substitute cultivation of *Ganoderma* come from a variety of sources, which contain lignin, cellulose, protein and minerals, and no heavy metal and pesticide residues. For example, sawdust of all kinds of broadleaf trees, crop straw (cotton stalk, soybean straw, wheat straw and corn cob), bagasse, reed straw and fruit shell (cottonseed shell and oak shell) can be mixed with an appropriate proportion of grain or the waste of grain processing (rice bran, wheat bran, corn flour, soybean flour and wheat flour). The sawdust of coniferous and aromatic trees cannot be used for the cultivation of *Ganoderma*. Material selection is necessary to overcome the inherent inadequacy of some materials. Sawdust should be screened before being used. Coarse sawdust and fine sawdust should be used together in the substrates. It is better to use more coarse sawdust to increase the permeability. The water content of the substrates should be controlled at 55%–60%, and the mixture should be kept in the reactor for 1–2 h to make it absorb water adequately. After mixing the substrates with water, it should be put into plastic bags in time. Because there is enough water in the substrates, all kinds of bacteria and fungi will be propagated very quickly which will affect the normal growth of *Ganoderma*. In high-temperature season, the materials mixed together should be bagged and sterilised timely.

Commonly used formulas are as follows [104]:

(i) Sawdust of broadleaf tree 79%, bran or rice bran 20%, gypsum powder 1%.
(ii) Cotton seed shell 89%, bran or rice bran 10%, gypsum powder 1%.
(iii) Corn cob 84%, bran or rice bran 10%, gypsum powder 1%.

In recent years, grass has been widely used as a substitute for the grass-cultivated technology. *Dicranopteris dichotoma*, *Miscanthus floridulu*, *Pennisetum purpureum* and *P. sinese* can be used for the grass cultivation of *Ganoderma*. There are two main methods for the artificial cultivation of grass: closing the hillsides for grass conservation and artificial cultivation. *Dicranopteris dichotoma* is widely distributed and difficult to cultivate. Therefore, it is more suitable for closing the hillsides for grass conservation. Fertilisation before rain can greatly increase the yield of *D. dichotoma*. Commonly used grasses, for example *P. purpureum*, *P. sinese*, *Themeda villosa* and *Neyraudia reynaudiana*, are generally cultivated artificially. *Pennisetum purpureum* and *P. sinese* can be harvested once or twice each year. Crush the grass and dry them after being harvested.

Commonly used formula are as follows [110]:

(i) 30% of *Dicranopteris dichotoma*, 48% of *Miscanthus floridulu*, 20% of bran, 2% of calcium carbonate.
(ii) 20% of *Dicranopteris dichotoma*, 58% of *Themeda villosa*, 20% of bran, 2% of calcium carbonate.
(iii) 39% of *Dicranopteris dichotoma*, 39% of *Miscanthus floridulu*, 20% of bran, 2% of calcium carbonate.
(iv) 26% of *Dicranopteris dichotoma*, 52% of *Phragmites communis*, 20% of bran, 2% of calcium carbonate.
(v) 83% of *Neyraudia reynaudiana*, 15% of bran, 2% of calcium carbonate.
(vi) 40% of *Dicranopteris dichotoma*, 43% of *Pennisetum purpureum*, 15% of bran, 2% of calcium carbonate.
(vii) 30% of *Dicranopteris dichotoma*, 53% of *Pennisetum sinese*, 15% of bran, 2% of calcium carbonate.

3. *Bagging*

There are two kinds of bagging methods: mechanical bagging and manual bagging. Mechanical bagging needs special machines, and the commonly used bagging machine are disc rotary punching type and spiral tube type. Bagged substrates should have a certain degree of tightness. Hyphae grow fast in the early stage and they are easily aging if the bagged is too loose. However, the ventilation is poor, hyphae grow slowly, the harvest of fruit bodies will be too late if the bag is too tight. The container to place the fungal bags must be smooth, no corners and no edges, to prevent puncturing plastic bags. Be careful when handling. After bagging, wipe up the residue attached to the ends of the bag. There is no fixed mode of bandaging the bags, making sure that the cold air is easy to overflow when sterilisation, and after sterilisation it can prevent the outside air flow into the bag directly.

4. *Sterilisation*

Start timing from bagging; the bags must be sterilised within 6 h. Polypropylene bags can be sterilised by autoclaving and atmospheric sterilisation.

Cooling: After sterilisation, the stove door can be opened when the temperature drops to 40–50 °C. At this time, the temperature inside the bag is higher than the outside, the air is flowing from the inside to the outside and the temperature is slightly higher, so it is not easy to be infected by bacteria or other fungi. The bags should be taken out of the stove and then sent to the cooling room for cooling. The transport tools must be cleaned beforehand. During the transportation, the fungi bag should be covered with a clean non-woven fabric or plastic film to prevent dust and other pollutants. The cooling chamber must be dry and clean, and no chemical pollution in the surrounding air.

5. *Inoculation*

When the bag is out of the stove, it can be inoculated when the temperature drops to 30 °C. Besides, the *Ganoderma* strain should not be inoculated when it is too cold. In order to reduce the contamination of bacteria and fungi, the amount of inoculum should be increased appropriately. The method and steps of inoculation are the same as wood-log cultivation. The strains must be in close contact with the substrate and the ends of the bag should be tightened in time.

2.4.2 *Management of Hypha Growth*

2.4.2.1 Wood-Log Cultivation

After inoculation, the lowest temperature in the greenhouse should be controlled between 20 and 22 °C and the hyphae will germinate in 2 or 3 days. Within a week, the temperature should be increased to 22–26 °C, which is beneficial to the growth of hyphae. The hyphae will join into pieces in about 7 days. The humidity should be maintained at 60%–65% in the initial stage and 65%–70% after 7 days. Hyphal growth begins with the formation of distinct hyphal circles on the surface of wood log and then gradually enters the xylem and pith to grow along vascular bundles. Generally, the key period of hyphae growth is 15 days after inoculation, and during this period ventilation, dehumidification and antibacteria are very important. With the growth of hyphae, the respiration rate gradually increases, and a large number of water droplets will be produced in the bag. At this time, it is necessary to strengthen ventilation and cooling. Open doors and windows for 1–2 h every afternoon. After 20 to 30 days of cultivation, the hyphae would grow on the whole surface of the wood. During the cultivation, the position of the fungi bags should be turned upside down, inside out and outside in. After 60 to 70 days of cultivation, the culture stage of hyphae can be completed. A few seriously polluted wood logs should be removed and re-inoculated. The excellent-quality wood logs are closely connected and difficult to separate. A small amount of yellowish-brown hyphae appears on the surface and will have elastic feeling when pressing the wood log. The weight of the wood log will reduce and the pale beige hyphae can be found in the xylem when the wood log is split.

2.4.2.2 Substitute Cultivation

After inoculation, the fungal bags should be placed in a ventilated, insulated and hygienic culture room, and try to avoid light. When inoculating in low-temperature season, the bags should be stacked centrally to increase the temperature and accelerate the colonisation of hyphae and then placed separately when the temperature rises. The temperature of the cultivation site should be controlled at 25–28 °C. In general, uncovering the straw curtain to ventilate can be used to control the temperature. During the hyphal growth, ventilation should be done in time to reduce air humidity. Generally, the ventilative frequency is about two or three times a day, about half an hour each time. The hyphae should be cultured in the dark before they are full in the bag, because the light will reduce the growth rate and make the fruit bodies form earlier.

In the hyphal culture stage, the turning should be done about every 10 days, and the position of the bags need be changed from top to bottom and from inside to outside. When turning, whether the fungal bags are contaminated by miscellaneous fungi or bacteria should be checked, and the contaminated bags should be picked out in time. After inoculation for 10–15 days, hyphae will ingest the substrate for about 4–5 cm from both ends of the bag inward. At this time, the tie rope at the neck of the bag should be removed to let a small amount of air into the bag. Under the suitable conditions, it takes about 30 days for the hyphae to grow up to the whole bag, and sometimes the fruit body primordia will appear when the hyphae reach 2/3.

2.4.3 Management of Fruiting

When the fungal bag is filled with hyphae or the surface of two ends appears primordium, it should enter into the period of fruit body management.

1. *Arrange Bags*

The bags are arranged ten layers in each row. Each layer of the bags is 1–2 cm apart. Two thin bamboo sticks are placed between the layers, which is conducive to ventilating and cooling, preventing the fruit body from collision and deformation during the growth process. The space between rows is 60–80 cm, which is convenient for manual operation management.

2. *Environmental Control During Fruit Body Growth*

The temperature should be controlled at 26–28 °C and the air relative humidity should be maintained at 90%–95%. Water management is advisable to a frequently small amount of water spraying, and keeping the ground moist. According to the local climate, the water spraying is usually twice a day. When spraying the ends of the bag, which should not have too much water, the primordia should not be sprayed directly. After spore powder is produced on the pore surface, it is not suitable to spray water on the pileus. Then ventilation should be strengthened; otherwise it will inhibit the differentiation of pileus or result in the deformity of the fruit body. The ventilation frequency and ventilation time should be flexible, generally two or three times a day and 1 h each time, and dry air should be avoided to blow on the fruit body directly. In a well-ventilated condition, fruit body will form early with short stipe, thick pileus and high yield. During the fruit body growth, sufficient light should be kept. In this case, the pileus will differentiate rapidly with lustrous colour, and the stipe will be short.

3. *Bud Pressing and Disbudding*

The temperature of the greenhouse should be kept at 26–28 °C, avoiding sharp temperature changes. If the temperature keeps below 18 °C or above 35 °C for a long time, the hyphae on the surface of the substrates will be chlorosis and the primordium will not differentiate. The suitable relative humidity for the forming of fruit body is 90%–95%. The low relative humidity can result in the death of the primordium. When the white fruit body primordia form at the two ends of the bag, tear the film of the ends, and roll it up. Keeping one or two primordia at the ends of each bag and pruning other primordium to concentrate nutrients, which will make a few pilei larger and thicker.

2.4.4 Fruit Body Harvest

When the pileus stops growing, the colour of the pileus is the same as the stipe, the edge of pileus has two or three layers of thickening line together with brown basidiospore dispersal, and the pore surface is beige to yellow, it can be harvested. To avoid basidiospore powder loss during harvest, the fruit body should be cut from the base of the stipe and pileus should not be touched or collided as much as possible. The fruit body cannot be washed with water; otherwise it will reduce the commercial value. After cutting the stipe and removing sediment and impurities, the single fruit body pileus should be put face up on the screen in the drying room. The baking temperature should be controlled as follows: 30–40 °C for 4 to 5 h in harvest day, raising to 55–60 °C for 1 or 2 h and drying to constant weight. When the water content of the fruit body reaches 11%–12%, it can be sealed and stored in a film bag. If basidiospore powder needs to be collected, it can be concentrated on kraft paper, newspaper or plastic film, making sure it is not mixed with dust or other sundries. Spore powder should also be dried, purified, wall-broken and refrigerated.

2.4.5 *Growing* Ganoderma *for Fruit Bodies and Spores*

2.4.5.1 Growing *Ganoderma* for Spores

'Hunong No 1' (*Ganoderma sichuanense*) is a cultivar that has the advantages of strong resistance to stress, easy collection of spore powder, high yield and good quality [111]. The inoculation of 'Hunong No. 1' is arranged in January to March, burying soil in April to May and collecting the spore powder in early August. The wood-log cultivation is generally used to cultivate 'Hunong No. 1', and the optimal wood log is Fagaceae trees, which can produce *Ganoderma* with good quality. 'Hunong No. 1' is mainly used for spore powder collection, and sleeving is the key technology. When the white growth rings of the fruit bodies disappear, colour turns brown, and a small amount of spore catapults, it is the time to sleeve. According to the size of fruit bodies, cylinders of different specifications are made of thin films (24 cm × 26 cm, 32 cm × 32 cm), and white cardboard should be cut into different sizes with diameter of 20–30 cm as cover board. Before sleeving, compact the soil on the ground. Lay a layer of thin film to separate the fruit body from the soil. Put the cylinder from the pileus to the bottom, tie the lower end of the cylinders to the stipe with a thin wire, then insert cardboard along the cylinder wall, staple the top with a stapler, and put on the cover board. In order to prevent sinking of the cover board on rainy days, which may cause mildew and death of the fruit body, the distance between the cover board and the pileus is greater than 5 cm. The cylinder should be kept at a distance from the edge of pileus because the spores would be unable to catapult if the thin films paste into the pore surface. When the sleeve steps are over, ridges need to be covered with film, being careful to prevent leaking of the film. During the whole process, it is better to control the temperature at 25 °C, and the relative humidity of the air at 75%–90%. It is suitable to harvest about 35 days after sleeving, at that time the pore surface becomes dark brown, and the spores stop catapulting. Take down the cover board and cylinder, and then cut the fruit body from the stipe. Carefully remove the spores that have accumulated, and then dry the spores on sunny days; the water content of spore powder should be controlled below 8% [111].

Besides 'Hunong No. 1', 'Liaolingzhi No. 2' (*G. sichuanense*) [112] and 'Xianzhi No. 2' (*G. sichuanense*) [113] both have high spore yield.

2.4.5.2 Growing *Ganoderma* for Fruit Bodies

'Longzhi No. 2' (*Ganoderma sichuanense*) is a cultivar not suitable for sporulation due to the weak ability to catapult spores, but it has the advantage of big fruit body, thick pileus and high yield. Fagaceae trees are suitable for wood-log cultivation of 'Longzhi No. 2'. Inoculation is arranged between January and early February, arranging and burying soil in April to May and harvesting the fruit body between middle July to middle September. In substitute cultivation, cultivation bags are usually made in late April, making an arrangement in June and harvesting at September.

'Longzhi No. 2' has high content of active ingredients; it is an excellent cultivar for cultivating fruit bodies and suitable for popularisation and application in production [114].

References

1. Karsten PA (1881) Enumeratio Boletinearum et Polyporearum Fennicarum, systemate novo dispositarum. Revue Mycol Paris 3:16–19
2. Moncalvo JM, Ryvarden L (1997) A nomenclatural study of the Ganodermataceae Donk. Fungiflora Norway Synop Fungorum 11:1–114
3. Audet S (2010) Essai de découpage systématique du genre Scutiger (Basidiomycota): *Albatrellopsis, Albatrellus, Polyporoletus, Scutiger* et description de six nouveaux genres. Mycotaxon, 111(1): 431–464
4. Anonymous (1995) Shen-Nong Materia Medica (102–200AD, reprinted in 1955). People's Hygiene Press, Beijing, p 514. (in Chinese)
5. Pegler DN (2002) Useful fungi of the world: the Ling-zhi – the mushroom of immortality. Mycologist 16:100–101
6. Elkhateeb WA, Zaghlol GM, El–Garawani IM, Ahmed EF et al (2018) *Ganoderma applanatum* secondary metabolites induced apoptosis through different pathways: *in vivo* and *in vitro* anticancer studies. Biomed Pharmacother 101:264–277
7. Hsu CC, Lin KY, Wang ZH, Lin WL et al (2008) Preventive effect of *Ganoderma amboinense* on acetaminophen-induced acute liver injury. Phytomedicine 15:946–950
8. Li CH, Chen PY, Chang UM, Kan LS et al (2005) Ganoderic acid X, a lanostanoid triterpene, inhibits topoisomerases and induces apoptosis of cancer cells. Life Sci 77:252–265
9. Jung SH, Lee YS, Shim SH, Lee S et al (2005) Inhibitory effects of *Ganoderma applanatum* on rat lens aldose reductase and sorbitol accumulation in streptozotocin-induced diabetic rat tissues. Phytother Res 19:477–480
10. Zhu K, Nie S, Li C, Xie M (2013) A newly identified polysaccharide from *Ganoderma atrum* attenuates hyperglycemia and hyperlipidemia. Int J Biol Macromol 57:142–150
11. Zhu KX, Nie SP, Tan LH, Li C et al (2016) A polysaccharide from *Ganoderma atrum* improves liver function in type 2 diabetic rats via antioxidant action and short-chain fatty acids excretion. J Agric Food Chem 64:1938–1944
12. El Dine RS, El Halawany AM, Ma CM, Hattori M (2009) Inhibition of the dimerization and active site of HIV-1 protease by secondary metabolites from the Vietnamese mushroom *Ganoderma colossum*. J Nat Prod 72:2019–2023
13. Jung M, Liermann JC, Opatz T, Erkel G (2011) Ganodermycin, a novel inhibitor of CXCL10 expression from *Ganoderma applanatum*. J Antibiot 64:683–686
14. Ko HH, Hung CF, Wang JP, Lin CN (2008) Anti-inflammatory triterpenoids and steroids from *Ganoderma lucidum* and *G. tsugae*. Phytochemistry 69:234–249
15. Li WJ, Li L, Zhen WY, Wang LF et al (2017) *Ganoderma atrum* polysaccharide ameliorates ROS generation and apoptosis in spleen and thymus of immunosuppressed mice. Food Chem Toxicol 99:199–208
16. Sliva D, Sedlak M, Slivova V, Valachovicova T et al (2003) Biologic activity of spores and dried powder from *Ganoderma lucidum* for the inhibition of highly invasive human breast and prostate cancer cells. J Altern Complement Med 9:491–497
17. Vazirian M, Dianat S, Manayi A, Ziari R et al (2014) Anti-inflammatory effect, total polysaccharide, total phenolics content and antioxidant activity of the aqueous extract of three basidiomycetes. Res J Pharm 1:13–19

18. Yan YM, Wang XL, Luo Q, Jiang LP et al (2015) Metabolites from the mushroom *Ganoderma lingzhi* as stimulators of neural stem cell proliferation. Phytochemistry 114:155–162
19. Yue GG, Fung KP, Leung PC, Lau CB (2008) Comparative studies on the immunomodulatory and antitumor activities of the different parts of fruiting body of *Ganoderma lucidum* and *Ganoderma* spores. Phytother Res 22:1282–1291
20. Zhou Y, Chen S, Ding R, Yao W et al (2014) Inflammatory modulation effect of Glycopeptide from *Ganoderma capense* (Lloyd) Teng. Mediat Inflam Article 2014:1–8
21. Bisko NA, Mitropolskaya NY (1999) Some biologically active substances from medicinal mushroom *Ganoderma lucidum*. Int J Med Mushrooms 5:301
22. Isaka M, Chinthanom P, Kongthong S, Srichomthong K et al (2013) Lanostane triterpenes from cultures of the Basidiomycete *Ganoderma orbiforme* BCC 22324. Phytochemistry 87:133–139
23. Mehta S (2014) Studies on genetic variability and bioactive molecules production by *Ganoderma* species. Shoolini University of Biotechnology and Management Sciences, Bajhol, Solan
24. Mothana RAA, Jansen R, Jülich WD, Lindequist U (2000) Ganomycin A and B, new antimicrobial farnesyl hydroquinones from the basidiomycete *Ganoderma pfeifferi*. J Nat Prod 63:416–418
25. Zhang XQ, Ip FC, Zhang DM, Chen LX et al (2011) Triterpenoids with neurotropic activity from *Ganoderma lucidum*. Nat Prod Res 25:1607–1613
26. Gao H, Chan E, Zhou F (2004) Immunomodulating activities of *Ganoderma*, a mushroom with medicinal properties. Food Rev Int 20:123–161
27. Chinese Pharmacopoeia Commission (1993) Pharmacopoeia of the People's Republic of China. China Medical Science Press, Beijing
28. Chang ST, Buswell JA (1999) *Ganoderma lucidum* (Curt. Fr.) P. Karst. (Aphyllophoromycetideae) – a mushrooming medicinal mushroom. Int J Med Mushrooms 1:139–146
29. Lai T, Gao Y, Zhou SF (2004) Global marketing of medicinal Ling Zhi mushroom *Ganoderma lucidum* (W.Curt:Fr.) Lloyd (Aphyllophoromycetideae) products and safety concerns. Int J Med Mushrooms 6:189–194
30. Singh SK, Doshi A, Pancholy A, Pathak R (2013) Biodiversity in wood–decay macro–fungi associated with declining arid zone trees of India as revealed by nuclear rDNA analysis. Eur J Plant Pathol 136(2):373–382
31. Jiang L (2015) *Ganoderma lucidum* (Reishi mushroom): potential application as health supplement and cosmeceutical ingredient. Glob J Res Anal 4:124–125
32. Chien CC, Tsai ML, Chen CC, Chang SJ et al (2008) Effects on tyrosinase activity by the extracts of Ganoderma lucidum and related mushrooms. Mycopathologia 166:117–120
33. Hyde KD, Bahkali AH, Moslem MA (2010) Fungi – an unusual source for cosmetics. Fungal Divers 43:1–9
34. Hapuarachchi KK, Elkhateeb WA, Karunarathna SC et al (2018) Current status of global *Ganoderma* cultivation, products, industry and market. Mycosphere 9(5):1025–1052
35. Wang XD, Zhang Y, Jiang HY, Luo Y, Zhang X, Li Y (2014) Cultivation of Zi-Zhi and bonsai production. Edible Fungi 36(04):51–52. (in Chinese)
36. Li QY, Zhong YY, Chen YX, Zhou WX, Zeng ZJ, Song B (2016) Research advances on the production technology of Ganoderma in China. Edible Fungi China 35(01):8–12+20. (in Chinese)
37. Patouillard N (1889) Le genre *Ganoderma*. Bull Soc Mycol France 5(2,3):64–80
38. Ames A (1913) A consideration of structure in relation to genera of the Polyporaceae. Ann Mycol 11(3):211–253
39. Bresadola G (1915) Basidiomycetes Philippinenses (Series III). Hedwigia 35:289–307
40. Imazeki R (1939) Studies on Ganoderma of Nippon. Bull Natl Sci Mus 1:29–52

41. Furtado JS (1965) Relation of microstructures to the taxonomy of the Ganodermoideae (Polyporaceae) with special reference to the structure of the cover of the pilear surface. Mycologia 57(4):588–611
42. Murrill WA (1950) The Polyporaceae of North America XI. A synopsis of the brown pileate species. Bull Torrey Bot Club 32(7):353–371
43. Murrill WA (1950) The Polyporaceae of North America XII. A synopsis of the white and bright-colored pileate species. Bull Torrey Bot Club 32(9):469–493
44. Murrill WA (1950) Tomophagus for Dendrophagus. Torreya 5:197
45. Steyaert RL (1972) Species of *Ganoderma* and related genera mainly of the Bogor and Leiden Herbaria. Persoonia 7(1):55–118
46. Ryvarden L (1991) Genera of polypores: nomenclature and taxonomy. Synop Fungorum 5:1–363
47. Donk MA (1993) Revision der Niederlandischen Homobasidiomycetae-Aphyllophoraceae 2. Meded Bot Mus Herb Rijksuniv Utrecht 9:1–278
48. Imazeki R (1943) Genera of polyporaceae of Nippon. Bull Tokyo Sci Mus 6:1–111
49. Imazeki R (1952) A contribution to the fungous flora of Dutch New Guinea. Bull Govt Forest Exp Stn Meguro 57:87–128
50. Steyaert RL (1980) Study of some *Ganoderma* species. Bull Jard Bot Natl Belg 50(1,2):135–186
51. Zhao JD, Xu LW, Zhang XQ (1979) Taxonomic studies of the subfamily Ganodermoideae of China. Acta Microbiol Sin 19(3):265–279. (in Chinese)
52. Jülich W (1981) Higher taxa of Basidiomycetes. Bibl Mycol 85:1–485
53. Corner EJH (1983) Ad Polyporaceae I – *Amauroderma* and *Ganoderma*. Beih Nova Hedwig 75:1–182
54. Ryvarden L (2004) Neotropical polypores Part 1. Synop Fungorum 19:1–229
55. Hong SG, Jung HS (2004) Phylogenetic analysis of *Ganoderma* based on nearly complete mitochondrial small-subunit ribosomal DNA sequences. Mycologia 96(4):742–755
56. Douanla-Meli C, Langer E (2009) *Ganoderma carocalcareus* sp. nov., with crumbly-friable context parasite to saprobe on *Anthocleista nobilis* and its phylogenetic relationship in *G. resinaceum* group. Mycol Prog 8(2):145–155
57. Kirk PM, Pr C, Minter DW, Sapers JA (2008) Dictionary of the fungi. C.A.B. International, Wallingford
58. Costa-Rezende DH, Robledo GL, Góes-Neto A et al (2017) Morphological reassessment and molecular phylogenetic analyses of *Amauroderma* s. lat. raised new perspectives in the generic classification of the Ganodermataceae family. Persoonia 39:254
59. Ma JX (1995) Noted Shen-Nong Materia Medica. People's Hygiene Press, Beijing. (in Chinese)
60. Su J et al (1981) Tang Materia Medica. Anhui Science & Technology Publishing House, Hefei, pp 659. (reprinted in 1981). (in Chinese)
61. Li SZ (1994) Compendium of Materia Medica. Ancient books of Traditional Chinese Medicine Press, Beijing, Beijing, pp 1552–1578. (in Chinese)
62. Tai FL (1979) Sylloge fungorum sinicorum. Science Press, Beijing
63. Zhao JD (1989) The Ganodermataceae in China. Bibliotheca Mycologica 132. J. Cramer, Berlin, p 176
64. Zhao JD, Zhang XQ (2000) Flora fungorum sinicorum 18, Ganodermataceae. Science Press, Beijing. (in Chinese)
65. Wu XL, Dai YC (2005) Coloured illustrations of Ganodermataceae of China. Science Press, Beijing, p 229. (in Chinese)
66. Wang XC, Xi RJ, Li Y, Wang DM, Yao YJ (2012) The species identity of the widely cultivated *Ganoderma*, '*G. lucidum*' (Ling-zhi), in China. PLoS ONE 7:e40857
67. Cao Y, Wu SH, Dai YC (2012) Species clarification of the prize medicinal Ganoderma mushroom "Lingzhi.". Fungal Divers 56:49–62

68. Yao YJ, Wang XC, Wang B (2013) Epitypification of *Ganoderma sichuanense* J.D. Zhao & X.Q. Zhang (Ganodermataceae). Taxon 62:1025–1031
69. Wu XL, Song B, Zhao YX, Wen TC, Huang SZ, Lin M (2013) Study on medicinal *Ganoderma* in China and discussion about name use of *Ganoderma* spp. Guizhou Sci 31(1):1–17. (in Chinese)
70. Bian YB (2017) Cultivation of edible mushroom. Higher Education Press, Beijing. (in Chinese)
71. Zhou LW, Cao Y, Wu SH, Vlasák J, Li DW, Li MJ, Dai YC (2015) Global diversity of the *Ganoderma lucidum* complex (Ganodermataceae, Polyporales) inferred from morphology and multilocus phylogeny. Phytochemistry 114:7–15
72. Ryvarden L (1985) Type studies in the Polyporaceae 17: species described by W. A. Murrill. Mycotaxon 23:169–198
73. Smith BJ, Sivasithamparam K (2003) Morphological studies of *Ganoderma* (Ganodermataceae) from the Australasian and Pacific regions. Aust Syst Bot 16(4):487–503
74. Zhao JD (1989) Studies on the taxonomy of Ganodermataceae in China IX. Subgen Ganoderma Sect Ganoderma Mycosystema 8(1):25–34. (in Chinese)
75. Moncalvo JM, Wang HF, Hseu RS (1995) Gene phylogeny of the *Ganoderma lucidum* complex based on ribosomal DNA sequences. Comparison with traditional taxonomic characters. Mycol Res 99:1489–1499
76. Adaskaveg JE, Gilbertson RL (1986) Cultural studies and genetics of sexuality of *Ganoderma lucidum* and *G. tsugae* in relation to the taxonomy of the *G. lucidum* complex. Mycologia 78:694–705
77. Hibbett DS, Donoghue MJ (1995) Progress toward a phylogenetic classification of the Polyporaceae through parsimony analysis of mitochondrial ribosomal DNA sequences. Can J Bot 73(S1):S853–S861
78. Hibbett DS, Thorn RG (2001) Basidiomycota: Homobasidiomycetes. The Mycota VII Part B. In: McLaughlin DJ, McLaughlin EG, Lemke PA (eds) Systematics and evolution. Springer, Berlin, pp 121–168
79. Moncalvo JM, Wang HH, Hseu RS (1995) Phylogenetic relationships in *Ganoderma* inferred from the internal transcribed spacers and 25S ribosomal DNA sequences. Mycologia 87:223–238
80. Bae SC, Lee SW, Kim HJ et al (1995) PCR amplification of ITS II region of rDNA for the classification of *Ganoderma* spp. RDA J Agric Sci 37:182–188
81. Thorn RG, Moncalvo JM, Reddy CA, Vilgalys R (2000) Phylogenetic analyses and the distribution of nematophagy support a monophyletic pleurotaceae within the polyphyletic pleurotoid-lentinoid fungi. Mycologia 92(2):241–252
82. Moncalvo JM, Vilgalys R, Redhead SA et al (2002) One hundred and seventeen clades of euagarics. Mol Phylogenet Evol 23(3):357–400
83. Hibbett DS, Pine EM, Langer E, Langer G, Donoghue MJ (1997) Evolution of gilled mushrooms and puffballs inferred from ribosomal DNA sequences. Proc Natl Acad Sci U S A 94(22):12002–12006
84. Richter C, Wittstein K, Kirk PM et al (2015) An assessment of the taxonomy and chemotaxonomy of *Ganoderma*. Fungal Divers 71(1):1–15
85. Moncalvo JM, Buchanan PK (2008) Molecular evidence for long distance dispersal across the southern hemisphere in the *Ganoderma applanatum-australe* species complex (Basidiomycota). Mycol Res 112:425–436
86. Hibbett DS, Grimaldi D, Donoghue MJ (1995) Cretaceous mushrooms in amber. Nature 377(6549):487–487
87. Thon MR, Royse DJ (1999) Partial β-tubulin gene sequences for evolutionary studies in the Basidiomycotina. Mycologia 91(3):468–474
88. Lan J, Xu JT, Wang QY (1998) Electrophoretic studies on estease and peroxidase isozymes in *Ganoderma* sp. Chin Pharm J 33(12):714–715

89. Pawlik A, Janusz G, Dębska I et al (2015) Genetic and metabolic intraspecific biodiversity of *Ganoderma lucidum*. Biomed Res Int 2015:1–13
90. Zheng LY, Jia DH, Fei XF, Luo X, Yang ZR (2009) An assessment of the genetic diversity within *Ganoderma* strains with AFLP and ITS PCR-RFLP. Microbiol Res 164(3):312–321
91. Mei ZQ, Yang LQ, Khan MA et al (2014) Genotyping of *Ganoderma* species by improved random amplifed polymorphic DNA (RAPD) and inter-simple sequence repeat (ISSR) analysis. Biochem Syst Ecol 56:40–48
92. Wang SH, Bai C, Fan J, Gao Y, Yang JF, Yang YJ (2003) A study on the RAPD analysis of *Ganoderma lucidum* and *Pleurotus ostreatus* protoplast fusant genome. Acta Edulis Fungi 10:1–5
93. Kwon O, Lee CS, Park YJ (2019) SNP and SCAR markers for specific discrimination of antler-shaped *Ganoderma lucidum*. Microorganisms 7(1):12
94. Xu MY, Tang CH, Zhang JS, Tang QJ, Yang Y, Jia W (2008) Development of SCAR markers based on SRAP and ISSR for rapid identifcation of *Ganoderma* strains. Mycosystema 27:707–717
95. Sun SJ, Gao W, Lin SQ, Zhu J, Xie BG, Lin ZB (2006) Analysis of genetic diversity in *Ganoderma* population with a novel molecular marker SRAP. Appl Microbiol Biotechnol 72(3):537–543
96. Gottlieb AM, Ferrer E, Wright JE (2000) rDNA analyses as an aid to the taxonomy of species of *Ganoderma*. Mycol Res 104(9):1033–1045
97. Smith BJ, Sivasithamparam K (2000) Internal transcribed spacer ribosomal DNA sequence of five species of *Ganoderma* from Australia. Mycol Res 104(8):943–951
98. Sotome K, Hattori T, Ota Y, To-anun C, Salleh B et al (2008) Phylogenetic relationships of *Polyporus* and morphologically allied genera. Mycologia 100(4):603–615
99. Hong SG, Jeong W, Jung HS (2002) Amplification of mitochondrial small subunit ribosomal DNA of polypores and its potential for phylogenetic analysis. Mycologia 94:823–833
100. Glen M, Bougher NL, Francis AA, Nigg SQ, Lee SS et al (2009) *Ganoderma* and *Amauroderma* species associated with root-rot disease of Acacia mangium plantation trees in Indonesia and Malaysia. Australas Plant Pathol 38:345–356
101. Wang DM, Wu SH, Su CH et al (2009) *Ganoderma multipileum*, the correct name for '*G. lucidum*' in tropical Asia. Bot Stu 50:451–458
102. Zhang X, Xu Z, Pei H et al (2017) Intraspecific variation and phylogenetic relationships are revealed by ITS1 secondary structure analysis and single-nucleotide polymorphism in *Ganoderma lucidum*. PLoS One 12(1):e0169042
103. Huang NL, Lin ZB, Chen GL (2010) Medicinal and edible fungi. Shanghai Scientific and Technological Literature Press, Shanghai
104. Wang DM (2005) Phlogenetic study of *Ganoderma* in China. Institute of Microbiology, Chinese Academy of Sciences, Beijing. (in Chinese)
105. Wang XC (2012) Phylogenetic study on Ganodermataceae Donk. Institute of Microbiology, Chinese Academy of Sciences, Beijing. (in Chinese)
106. Pilotti CA, Sanderson FR, Aitken AB (2003) Genetic structure of a population of *Ganoderma boninense* on oil palm. Plant Pathol 52(4):455–463
107. Pilotti CA, Sanderson FR, Aitken AB (2002) Sexuality and interactions of monokaryotic and dikaryotic mycelia of *Ganoderma boninense*. Mycol Res 106(11):1315–1322
108. Yu WY (2017) Study on introduction and cultivation techniques of *Ganoderma lucidum*. Zhejiang University, Hangzhou. (in Chinese)
109. Chen RY, Kang J, Du GH (2016) Construction of the quality control system of *Ganoderma* products. Edib Med Mushrooms 24(6):339–344
110. Lin ZB (2007) Modern research on Ganoderma. Peking University Medical Press, Beijing. (in Chinese)
111. He JF, Li ZQ, Zhang LM (2011) Breeding of *Ganoderma* cultivar "Hunong 1" and its characteristics. Edib Fungi 33(05):21–22. (in Chinese)

112. Liu YY, Song Y, Liu N, Liu JJ (2018) Traits comparison between new *Ganoderma* cultivar 'Liaolingzhi 2' and main cultivars in Liaoning. North Hortic 04:165–168. (in Chinese)
113. Li JM, Xu J, Wang XJ, Zhang GL, Li MY, Shao QS, Pan QX (2017) Breeding of *Ganoderma* var. "Xianzhi 2 and its characteristics". Mod Chinese Med 19(03):324–345. (in Chinese)
114. Li ZQ, Cai WM, Han HP, Ye XJ, Jin QL, Li C, Li XJ (2016) Breeding of *Ganoderma* cultivar "Longzhi 2" and its characteristics. Edib Med Mushrooms 24(02):118–119. (in Chinese)

Chapter 3
Chemical Components of *Ganoderma*

Ting Gong, Renyi Yan, Jie Kang, and Ruoyun Chen

Abstract So far, more than 20 species of *Ganoderma* (Lingzhi) fungi have been studied, including *G. lucidum*, *G. sinense*, *G. japonicum*, *G. capense*, *G. australe*, *G. tsugae*, *G. applanatum*, *G. tropicus*, *G. boniense*, *G. duropora*, *G. resinaceum*, *G. theaecolum*, *G. cochlear*, *G. atrum*, *G. formosanum*, *G. boninense*, *G. colossum*, *G. concinna*, *G. amboinense*, *G. pfeifferi*, and *G. orbiforme*. Over 600 compounds were isolated and identified from the genus *Ganoderma*. The components in *Ganoderma* contained triterpenes, meroterpenoids, steroids, alkaloids, nucleosides, nucleobases, and polysaccharides, in which triterpenes were the main compounds, and more than 300 triterpenes obtained from it. With the use of new technology of separation and purification, the other types of compounds, such as meroterpenoids, alkaloids, nucleosides, and polysaccharides were reported in recent years. Herein, the structural classifications and characteristics and separation methods were summarized.

Keywords Ganoderma · Triterpenes · Meroterpenoids · Steroids · Alkaloids and nucleosides · Polysaccharides

3.1 Triterpenoids

Up to now, over 300 triterpenoids have been isolated from the genus *Ganoderma*. Herein, the characteristics of triterpenoids were summarized. The advantages and disadvantages of extraction for *Ganoderma* triterpenoids, including organic solvent extraction, ultrasound-assisted extraction, microwave-assisted extraction, and supercritical fluid extraction, were compared.

T. Gong · R. Yan · J. Kang · R. Chen (✉)
Department of Natural Product Chemistry, Institute of Materia Medica, Chinese Academy of Medical Sciences & Peking Union Medical College, Beijing, China
e-mail: rych@imm.ac.cn

Z. Lin, B. Yang (eds.), *Ganoderma and Health*, Advances in Experimental Medicine and Biology 1181, https://doi.org/10.1007/978-981-13-9867-4_3

3.1.1 *Chemical Structures of* Ganoderma *Triterpenoids*

The structures of triterpenoids isolated from the genus *Ganoderma* are complicated. These compounds include tetracyclic (lanostane carbon skeleton) and pentacyclic triterpenoids. According to the numbers of carbons, the triterpenoids can be divided into three groups, which are C_{30}, C_{27}, and C_{24}. They can also be grouped into triterpenoid acids, triterpenoid alcohols, and triterpenoid lactones based on the substituting groups. The compounds were arranged for convenience in the alphabetic order of the names, and detailed information was shown in Table 3.1 and Fig. 3.1.

3.1.2 *Extraction of* Ganoderma *Triterpenoids*

3.1.2.1 Organic Solvent Extraction

Ethanol and methanol are commonly used to extract the triterpenoids from *Ganoderma* fungi [93]. Organic solvent extraction is a traditional extraction method, but it is a time-consuming process and requires large amounts of solvents. Besides, the efficiency of extraction is low. However, it is still a common traditional extraction method widely used in isolation of *Ganoderma* triterpenoids in most laboratories.

3.1.2.2 Ultrasound-Assisted Extraction

Ultrasound-assisted extraction is an extensively used alternative extraction method for traditional organic solvent extraction, featuring the lower extraction times, lower solvent consumption, and higher extraction efficiency.

Gong XF et al. [94] found that using ultrasound-assisted extraction could get higher extraction efficiency of *Ganoderma* triterpenes (1.43%) than using organic solvent extraction (0.9294%) under the same condition [94].

3.1.2.3 Microwave-Assisted Extraction

Compared with traditional organic solvent extraction, microwave-assisted extraction takes much less time to get the better result.

Chen Yi et al. (2007) used the method of microwave-assisted extraction to extract total triterpenoid saponins from *G. atrum* for the first time. The whole extraction took only 5 min, and the extraction efficiency of triterpenoid saponins was 0.968%. Optimal condition of microwave-assisted extraction for *Ganoderma* fungi is as follows: extraction time, 5 min; extraction temperature, 90 °C; extraction solvent, 95% ethanol [95].

Table 3.1 Chemical names and molecular formulas of *Ganoderma* triterpenoids

No	Chemical names	Formulas	Ref
1	3*α*-acetoxy-5*α*-lanosta-8, 24-dien-21-oic acid ester *β*-D-glucoside	$C_{38}H_{60}O_9$	[1]
2	3*α*-(3-hydroxy-5-methoxy-3-methyl-1,5-dioxopentyl-oxy)-24-methylene-5*α*-lanost-8-en-21-oic acid	$C_{38}H_{60}O_7$	[2]
3	3-epipachymic acid (3*α*-acetoxy-16α-hydroxy-24-methylene-5*α*-lanost-8-en-21-oic acid	$C_{33}H_{52}O_5$	[2]
4	3*β*-acetoxy-5-lanosta-8,24(24′)-dien-21-oic acid	$C_{33}H_{52}O_4$	[3]
5	3*β*-hydroxy-15*α*-acetoxy-5*α*-lanosta-7,9(11),24-trien-26-al	$C_{32}H_{48}O_4$	[3]
6	3*β*-hydroxy-7,22-dioxo-5*α*-lanosta-8,24-dien-21-oic acid	$C_{30}H_{44}O_5$	[3]
7	3*β*,7*β*-dihydroxy-12*β*-acetoxy-11,15,23-trioxo-5*α*-lanosta-8-en-26-oic acid methyl ester	$C_{33}H_{48}O_9$	[4]
8	3*β*,12*β*-dihydroxy-7,11,15,23-tetraoxo-lanost-8,20-dien-26-oic acid	$C_{30}H_{40}O_8$	[5]
9	3*β*,15*β*-dihydroxy-7,11,23-trioxo-lanost-8,16-dien-26-oic acid methyl ester	$C_{31}H_{44}O_7$	[6]
10	3*α*,15*α*-diacetoxy-22*α*- hydroxy-lanosta-7,9(11),24-trien-26-oic acid	$C_{34}H_{50}O_7$	[7]
11	3*β*,15*α*-diacetoxy-22*β*-hydroxy-lanosta-7,9(11),24-trien-26-oic acid	$C_{34}H_{50}O_7$	[7]
12	3,15-diacetoxy-lanosta-8,24-dien-26-oic acid	$C_{35}H_{54}O_6$	[8]
13	3*α*,22*β*-diacetoxy-7*α*- hydroxyl-5*α*-lanost-8,24*E*-dien-26-oic acid	$C_{34}H_{52}O_7$	[9]
14	3*β*,7*β*,12*β*-trihydroxy-11,15,23-trioxo-lanost-8,20-dien-26-oic acid	$C_{30}H_{42}O_8$	[5]
15	3*β*,7*β*,15*α*-trihydroxy-4-(hydroxymethyl)-11,23-dioxo-lanost-8-en-26-oic acid	$C_{30}H_{46}O_9$	[5]
16	3*β*,7*β*,15*β*-trihydroxy-11,23-dioxo-lanost-8,16-dien-26-oic acid methyl ester	$C_{31}H_{46}O_7$	[6]
17	3*β*,7*β*,15*β*-trihydroxy-11,23-dioxo-lanost-8,16-dien-26-oic acid	$C_{30}H_{44}O_7$	[6]
18	3,12*β*,15-triacetoxy-5-lanosta-7,9(11),24-trien-26-oic acid	$C_{36}H_{52}O_8$	[10]
19	3*β*,15*α*,22*β*-trihydroxylanost-7,9 (11),24- trien-26-oic acid	$C_{30}H_{46}O_5$	[7]
20	3*β*,7*β*,15,24-tetrahydroxy-11,23-dioxo-lanost-8-en-26-oic acid	$C_{30}H_{46}O_8$	[11]
21	3*β*,7*β*,15,28-tetrahydroxy-11,23-dioxo-lanost-8,16-dien-26-oic acid	$C_{30}H_{44}O_8$	[11]
22	4,4,14*α*-trimethyl-3,7-dioxo-5*α*-achol-8-en-24-oic acid	$C_{27}H_{40}O_4$	[4]
23	5*α*-lanosta-7,9(11), 24-triene-15*α*-26- dihydroxy-3-one	$C_{30}H_{46}O_3$	[12]
24	5*α*-lanosta-7,9(11), 24-triene-3*β*-hydroxy-26-al	$C_{30}H_{46}O_2$	[12]
25	5-lanosta-8,24-diene-26,27-dihydroxy-3,7-dione	$C_{30}H_{46}O_3$	[10]
26	8*α*,9*α*-epoxy-3,7,11,15,25-pentaoxo-5*α*-lanosta-26-oic acid	$C_{30}H_{40}O_8$	[13]
27	8*α*,9*α*-epoxy-4,4, 14*α*-trimethyl-3,7,11,15,20-pentaoxo-5*α*-pregnane	$C_{24}H_{30}O_6$	[12]
28	8*β*,9*α*-dihydroganoderic acid J	$C_{30}H_{44}O_7$	[12]
29	11*α*-hydroxy-3,7-dioxo-5*α*-lanosta-8, 24(*E*)-dien-26-oic acid	$C_{30}H_{44}O_5$	[4]
30	11*β*-hydroxy-3,7-dioxo-5*α*-lanosta-8, 24(*E*)-dien-26-oic acid	$C_{30}H_{44}O_5$	[4]
31	12-acetoxy-7-hydroxy-3,11,15,23-tetraoxo-5-lanosta-8,20-dien-26-oic acid ethyl ester	$C_{34}H_{48}O_9$	[14]
32	12*β*-acetoxy-3*β*-hydroxy-7,11,15,23-tetraoxo-lanost-8,20*E*-diene-26-oic acid	$C_{32}H_{42}O_9$	[15]
33	12*β*-acetoxy-3,7,11,15, 23-pentaoxo-5*α*-lanosta-8-en-26-oic acid ethyl ester	$C_{34}H_{46}O_9$	[4]
34	12*β*-acetoxy-3,7,11,15,23-pentaoxolanost-8-en-26-oic acid butyl ester	$C_{36}H_{50}O_9$	[16]

(continued)

Table 3.1 (continued)

35	12*β*-acetoxy-3,7,11,15,23-pentaoxo-lanost-8,20-dien-26-oic acid	$C_{30}H_{42}O_9$	[5]
36	12*β*-acetoxy-3*β*,7*β*- dihydroxy-11,15,23- trioxolanost-8-en-26-oic acid butyl ester	$C_{36}H_{54}O_9$	[16]
37	12*β*-acetoxy-7*β*-hydroxy-3,11,15, 23-tetraoxo-5*α*- lanosta-8, 20-dien-26-oic acid ethyl ester	$C_{34}H_{48}O_9$	[14]
38	15*α*-acetoxy-3*α*-hydroxylanosta-8,24-dien-26-oic acid	$C_{32}H_{50}O_5$	[17]
39	15*α*-acetoxy-5*α*-lanosta-7,9(11),24-trien-3*β*,26-diol	$C_{32}H_{50}O_4$	[3]
40	15*α*-hydroxy-3,11,23-trioxo-lanost-8,20-dien-26-oic acid	$C_{30}H_{42}O_6$	[5]
41	16,20-dihydroxy-3,23-dioxo-5-lanosta-6,8-dien-26-oic acid	$C_{30}H_{44}O_6$	[3]
42	16*α*,26-dihydroxylanosta-8,24-dien-3-one	$C_{30}H_{48}O_3$	[18]
43	20-(21)dehydrolucidenic acid N	$C_{27}H_{38}O_6$	[19]
44	20(21)-dehydrolucidenic acid A	$C_{27}H_{36}O_6$	[20]
45	20-hydroxy-3,12,15,23-tetraoxolanosta-7,9(11),16-trien-26-oic acid	$C_{30}H_{38}O_7$	[21]
46	20-hydroxyganoderic acid AM1	$C_{30}H_{42}O_8$	[22]
47	20-hydroxylganoderic acid G	$C_{30}H_{44}O_9$	[12]
48	20-hydroxylucidenic acid A	$C_{27}H_{38}O_7$	[19]
49	20-hydroxylucideric acid D2	$C_{29}H_{38}O_9$	[20]
50	20-hydroxylucideric acid E2	$C_{29}H_{40}O_9$	[20]
51	20-hydroxylucideric acid F	$C_{27}H_{36}O_7$	[20]
52	20-hydroxylucideric acid N	$C_{27}H_{40}O_7$	[20]
53	20-hydroxylucideric acid P	$C_{29}H_{42}O_9$	[20]
54	22*β*-acetoxy-3*α*,15*α*-dihydroxy-lanosta-7,9(11), 24-trien-26-oic acid	$C_{32}H_{48}O_6$	[23]
55	22*β*-acetoxy-3*β*,15*α*-dihydroxy-lanosta-7,9(11), 24-trien-26-oic acid	$C_{32}H_{48}O_6$	[23]
56	(22*S*,24*E*)-15*α*,22-diacetoxy-3-oxolanosta-8,24-dien-26-oic acid	$C_{34}H_{50}O_7$	[24]
57	(22*S*,24*E*)-15*α*,22-diacetoxy-3*β*-hydroxylanosta-8,24-dien-26-oic acid	$C_{34}H_{52}O_7$	[24]
58	(22*S*,24*E*)-22-acetoxy-3,7-dioxolanosta-8,24-dien-26-oic acid	$C_{32}H_{46}O_6$	[24]
59	(22*S*,24*E*)-22-acetoxy-3*β*-hydroxylanosta-7,9(11),24-trien-26-oic acid	$C_{34}H_{50}O_6$	[24]
60	(22*S*,24*E*)-3*α*,22-diacetoxy-7-oxolanosta-8,24-dien-26-oic acid	$C_{34}H_{50}O_7$	[24]
61	(22*S*,24*E*)-3*β*,15*α*,22-triacetoxylanosta-8,24-dien-26-oic acid	$C_{36}H_{54}O_8$	[24]
62	(22*S*,24*E*)-3*β*,22-diacetoxy-7*α*-methoxylanosta-8,24-dien-26-oic acid	$C_{35}H_{54}O_7$	[24]
63	(22*S*,24*E*)-3*β*,22-diacetoxylanosta-7,9(11),24-trien-26-oic acid	$C_{34}H_{50}O_6$	[24]
64	(22*S*,24*E*)-7*α*-hydroxy-3*β*,15*α*,22-triacetoxylanosta-8,24-dien-26-oic acid	$C_{36}H_{54}O_9$	[24]
65	(22*S*,24*E*)-7α-methoxy-3*β*,15*α*,22-triacetoxylanosta-8,24-dien-26-oic acid	$C_{37}H_{56}O_9$	[24]
66	(22*Z*,24*Z*)-13-hydroxy-3-oxo-14(13 → 12) abeo-lanosta-8,22,24-trien-26,23-olide	$C_{30}H_{42}O_4$	[25]
67	23(*S*)-hydroxy-3,7,11,15- tetraoxo-lanost-8,24*E*- diene-26-oic acid	$C_{30}H_{40}O_7$	[15]
68	24(*R*)-tirucalla-7,9(11),25-triene-3,24,27-triol	$C_{30}H_{48}O_4$	[26]
69	24(*S*)-tirucalla-7,9(11),25-triene-3,24,27-triol	$C_{30}H_{48}O_4$	[26]
70	(24*E*)-3*β*,15*α*-diacetoxy-7*α*-hydroxylanosta-8,24-dien-26-oic acid	$C_{34}H_{52}O_7$	[24]
71	(24*E*)-3*β*-acetoxy-15*α*-hydroxy-7*α*-methoxylanosta-8,24-dien-26-oic acid	$C_{33}H_{52}O_6$	[24]
72	(24*E*)-3*β*-acetoxy-7*α*-hydroxylanosta-8,24-dien-26-oic acid	$C_{32}H_{50}O_5$	[24]

(continued)

Table 3.1 (continued)

73	(24*E*)-3*β*-acetoxylanosta-7,9(11),24-trien-26-oic acid					$C_{32}H_{48}O_4$	[24]
74	(24*E*)-7*α*-acetoxy-15-hydroxy-3-oxolanosta-8,24-dien-26-oic acid					$C_{32}H_{48}O_6$	[24]
75	(24*E*)-7*α*-methoxy-3-oxolanosta-8,24-dien-26-oic acid					$C_{31}H_{48}O_4$	[24]
76	(24*E*)-9,11-epoxy-3*β*-hydroxylanosta-7,24-dien-26-al					$C_{30}H_{46}O_3$	[25]
77	(24*S*)-3-oxo-7,24,25-trihydroxylanosta-8-ene					$C_{30}H_{50}O_4$	[27]
78	26,27-dihydroxy-5α-lanosta-7,9(11)24-triene-3,22-dione					$C_{30}H_{44}O_4$	[28]
No	**Chemical name**	**Formula**	**Ref**	**No**	**Chemical name**	**Formula**	**Ref**
79	Applanoxidic acid G	$C_{30}H_{40}O_8$	[29]	**80**	Butyl lucidenate D2	$C_{33}H_{48}O_8$	[30]
81	Butyl lucidenate E2	$C_{33}H_{46}O_8$	[30]	**82**	Butyl lucidenate P	$C_{33}H_{50}O_8$	[30]
83	Butyl lucidenate Q	$C_{31}H_{48}O_6$	[30]	**84**	Cochlate A	$C_{28}H_{40}O_6$	[31]
85	Cochlate B	$C_{28}H_{40}O_6$	[31]	**86**	Cochlate C	$C_{28}H_{38}O_7$	[32]
87	Cochlearic acid A	$C_{27}H_{38}O_6$	[32]	**88**	Cochlearic acid B	$C_{30}H_{46}O_3$	[32]
89	Colossolactone A	$C_{32}H_{52}O_5$	[33]	**90**	Colossolactone B	$C_{32}H_{48}O_5$	[33]
91	Colossolactone C	$C_{32}H_{46}O_6$	[33]	**92**	Colossolactone D	$C_{32}H_{46}O_6$	[33]
93	Colossolactone E	$C_{32}H_{42}O_6$	[33]	**94**	Colossolactone F	$C_{32}H_{42}O_7$	[33]
95	Colossolactone G	$C_{32}H_{42}O_7$	[29, 33]	**96**	Compounds B8	$C_{30}H_{44}O_7$	[34]
97	Compounds B9	$C_{30}H_{46}O_7$	[34]	**98**	Daqingshone A	$C_{24}H_{30}O_5$	[35]
99	Daqingshone B	$C_{27}H_{40}O_6$	[35]	**100**	Diacetate ganoderiol C	$C_{36}H_{58}O_7$	[36]
101	Diacetate ganoderiol D	$C_{34}H_{52}O_7$	[36]	**102**	Epoxyganoderiol A	$C_{30}H_{48}O_4$	[37]
103	Epoxyganoderiol B	$C_{30}H_{46}O_3$	[37]	**104**	Epoxyganoderiol C	$C_{30}H_{48}O_3$	[37]
105	Ermanondiol	$C_{30}H_{48}O_3$	[38]	**106**	Ethyl ganoderenate D	$C_{32}H_{44}O_7$	[14]
107	Ethyl lucidenate A	$C_{29}H_{42}O_6$	[39]	**108**	Fornicatin A	$C_{27}H_{40}O_7$	[40]
109	Fornicatin B	$C_{27}H_{40}O_6$	[40]	**110**	Fornicatin D	$C_{28}H_{42}O_6$	[31]
111	Fornicatin E	$C_{28}H_{42}O_6$	[31]	**112**	Fornicatin F	$C_{29}H_{44}O_6$	[31]
113	Fornicatin G	$C_{29}H_{44}O_6$	[41]	**114**	Fornicatin H	$C_{29}H_{44}O_7$	[41]
115	Ganoboninketal D	$C_{31}H_{44}O_8$	[27]	**116**	Ganoboninketal A	$C_{32}H_{46}O_7$	[42]
117	Ganoboninketal B	$C_{30}H_{42}O_6$	[42]	**118**	Ganoboninketal C	$C_{33}H_{48}O_8$	[42]
119	Ganoboninone A	$C_{29}H_{38}O_8$	[43]	**120**	Ganoboninone B	$C_{29}H_{40}O_8$	[43]
121	Ganoboninone C	$C_{29}H_{40}O_8$	[43]	**122**	Ganoboninone D	$C_{30}H_{42}O_8$	[43]
123	Ganoboninone E	$C_{29}H_{40}O_7$	[43]	**124**	Ganoboninone F	$C_{32}H_{46}O_7$	[43]
125	Ganochlearic acid A	$C_{24}H_{34}O_5$	[32]	**126**	Ganoderal A	$C_{30}H_{44}O_2$	[44]
127	Ganoderal B	$C_{30}H_{46}O_3$	[37]	**128**	Ganodercochlearin A	$C_{30}H_{48}O_3$	[45]
129	Ganodercochlearin B	$C_{30}H_{48}O_3$	[45]	**130**	Ganodercochlearin C	$C_{31}H_{50}O_3$	[45]
131	Ganodercochlearin D	$C_{30}H_{48}O_3$	[32]	**132**	Ganodercochlearin E	$C_{30}H_{50}O_4$	[32]
133	Ganodercochlearin F	$C_{30}H_{48}O_3$	[32]	**134**	Ganodercochlearin G	$C_{31}H_{50}O_3$	[32]
135	Ganodercochlearin H	$C_{32}H_{50}O_3$	[32]	**136**	Ganodercochlearin I	$C_{30}H_{46}O_3$	[32]
137	Ganodercochlearin J	$C_{30}H_{46}O_3$	[32]	**138**	Ganodercochlearin K	$C_{31}H_{52}O_3$	[32]
139	Ganoderenic acid A	$C_{30}H_{42}O_7$	[38]	**140**	Ganoderenic acid AM1	$C_{30}H_{40}O_7$	[22]
141	Ganoderenic acid B	$C_{30}H_{42}O_7$	[38]	**142**	Ganoderenic acid C	$C_{30}H_{44}O_7$	[38]
143	Ganoderenic acid D	$C_{30}H_{40}O_7$	[38]	**144**	Ganoderense A	$C_{30}H_{40}O_7$	[46]
145	Ganoderenses B	$C_{30}H_{42}O_7$	[46]	**146**	Ganoderense C	$C_{31}H_{42}O_7$	[46]
147	Ganoderenses D	$C_{31}H_{44}O_8$	[46]	**148**	Ganoderense E	$C_{31}H_{48}O_8$	[46]

(continued)

Table 3.1 (continued)

149	Ganoderesin C	$C_{30}H_{42}O_7$	[22]	**150**	Ganoderic acid AM1	$C_{30}H_{42}O_7$	[12]
151	Ganoderic acid A	$C_{30}H_{44}O_7$	[38]	**152**	Ganoderic acid B	$C_{30}H_{44}O_7$	[38]
153	Ganoderic acid C	$C_{30}H_{42}O_7$	[47]	**154**	Ganoderic acid C2	$C_{30}H_{46}O_7$	[34]
155	Ganoderic acid C6	$C_{30}H_{42}O_8$	[38]	**156**	Ganoderic acid Df	$C_{30}H_{44}O_7$	[48]
157	Ganoderic acid DM	$C_{30}H_{44}O_4$	[49]	**158**	Ganoderic acid GS-1	$C_{30}H_{42}O_6$	[19]
159	Ganoderic acid GS-2	$C_{30}H_{44}O_6$	[19]	**160**	Ganoderic acid GS-3	$C_{32}H_{46}O_8$	[19]
161	Ganoderic acid K	$C_{32}H_{46}O_9$	[44]	**162**	Ganoderic acid LM2	$C_{30}H_{42}O_7$	[50]
163	Ganoderic acid Ma	$C_{34}H_{52}O_7$	[51]	**164**	Ganoderic acid Mb	$C_{36}H_{54}O_9$	[51]
165	Ganoderic acid Mc	$C_{36}H_{54}O_9$	[51]	**166**	Ganoderic acid Md	$C_{35}H_{54}O_7$	[51]
167	Ganoderic acid Me	$C_{34}H_{50}O_6$	[51]	**168**	Ganoderic acid Mf	$C_{32}H_{48}O_5$	[52]
169	Ganoderic acid Mg	$C_{35}H_{54}O_8$	[53]	**170**	Ganoderic acid Mh	$C_{34}H_{52}O_8$	[53]
171	Ganoderic acid Mi	$C_{33}H_{52}O_6$	[53]	**172**	Ganoderic acid Mj	$C_{33}H_{52}O_6$	[53]
173	Ganoderic acid P	$C_{34}H_{50}O_7$	[8]	**174**	Ganoderic acid Q	$C_{34}H_{50}O_7$	[53]
175	Ganoderic acid R	$C_{34}H_{50}O_6$	[54]	**176**	Ganoderic acid S1	$C_{30}H_{44}O_3$	[44]
177	Ganoderic acid S2	$C_{33}H_{48}O_5$	[55]	**178**	Ganoderic acid T	$C_{36}H_{52}O_8$	[55]
179	Ganoderic acid U	$C_{30}H_{48}O_4$	[52]	**180**	Ganoderic acid V	$C_{32}H_{48}O_6$	[52]
181	Ganoderic acid W	$C_{34}H_{52}O_7$	[52]	**182**	Ganoderic acid X	$C_{33}H_{50}O_5$	[56]
183	Ganoderic acid XL1	$C_{30}H_{46}O_7$	[22]	**184**	Ganoderic acid XL2	$C_{30}H_{46}O_7$	[22]
185	Ganoderic acid XL3	$C_{30}H_{46}O_5$	[57]	**186**	Ganoderic acid XL4	$C_{30}H_{40}O_7$	[57]
187	Ganoderic acid XL5	$C_{31}H_{48}O_8$	[57]	**188**	Ganoderic acid Y	$C_{31}H_{46}O_3$	[58]
189	Ganoderic acid Z	$C_{30}H_{48}O_3$	[52]	**190**	Ganoderic acid α	$C_{32}H_{46}O_9$	[59]
191	Ganoderic acid β	$C_{32}H_{46}O_6$	[60]	**192**	Ganoderic acid γ	$C_{30}H_{44}O_7$	[61]
193	Ganoderic acid δ	$C_{30}H_{44}O_7$	[61]	**194**	Ganoderic acid ε	$C_{30}H_{44}O_7$	[50]
195	Ganoderic acid ζ	$C_{30}H_{42}O_7$	[61]	**196**	Ganoderic acid η	$C_{30}H_{44}O_8$	[61]
197	Ganoderic acid θ	$C_{30}H_{42}O_8$	[61]	**198**	Ganoderiol A	$C_{30}H_{50}O_4$	[62]
199	Ganoderiol B	$C_{30}H_{46}O_4$	[62]	**200**	Ganoderiol F	$C_{30}H_{46}O_3$	[36]
201	Ganoderiol H	$C_{30}H_{50}O_5$	[36]	**202**	Ganoderiol I	$C_{31}H_{50}O_5$	[36]
203	Ganoderlactone A	$C_{27}H_{36}O_7$	[63]	**204**	Ganoderlactone B	$C_{27}H_{34}O_6$	[63]
205	Ganoderlactone C	$C_{29}H_{36}O_8$	[63]	**206**	Ganoderlactone D	$C_{27}H_{38}O_7$	[63]
207	Ganoderlactone E	$C_{27}H_{36}O_7$	[63]	**208**	Ganodermadiol	$C_{30}H_{48}O_2$	[29]
209	Ganodermalactone A	$C_{30}H_{40}O_3$	[64]	**210**	Ganodermalactone B	$C_{32}H_{46}O_6$	[64]
211	Ganodermalactone C	$C_{30}H_{44}O_6$	[64]	**212**	Ganodermalactone D	$C_{33}H_{48}O_8$	[64]
213	Ganodermalactone E	$C_{30}H_{46}O_4$	[64]	**214**	Ganodermalactone F	$C_{30}H_{38}O_5$	[64]
215	Ganodermalactone G	$C_{30}H_{36}O_6$	[64]	**216**	Ganodermanontriol	$C_{30}H_{48}O_4$	[62]
217	Ganodermatriol	$C_{30}H_{48}O_3$	[65]	**218**	Ganodernoid A	$C_{25}H_{32}O_6$	[63]
219	Ganodernoid B	$C_{25}H_{34}O_6$	[63]	**220**	Ganodernoid C	$C_{28}H_{36}O_6$	[63]
221	Ganodernoid D	$C_{32}H_{40}O_9$	[63]	**222**	Ganodernoid E	$C_{31}H_{46}O_8$	[63]
223	Ganodernoid F	$C_{31}H_{44}O_8$	[63]	**224**	Ganodernoid G	$C_{32}H_{44}O_9$	[63]
225	Ganoderol A	$C_{30}H_{44}O_2$	[44]	**226**	Ganoderol B	$C_{30}H_{46}O_3$	[37]
227	Ganoduritriol A	$C_{30}H_{48}O_3$	[66]	**228**	Ganoduritriol B	$C_{30}H_{50}O_3$	[66]
229	Ganolactone B	$C_{27}H_{38}O_6$	[67]	**230**	Ganolactone	$C_{27}H_{36}O_6$	[65]
231	Ganoleuconin A	$C_{30}H_{44}O_6$	[68]	**232**	Ganoleuconin B	$C_{30}H_{44}O_7$	[68]
233	Ganoleuconin C	$C_{32}H_{46}O_8$	[68]	**234**	Ganoleuconin D	$C_{32}H_{44}O_9$	[68]

(continued)

Table 3.1 (continued)

235	Ganoleuconin E	$C_{30}H_{42}O_8$	[68]	**236**	Ganoleuconin F	$C_{32}H_{42}O_9$	[68]
237	Ganoleuconin G	$C_{30}H_{46}O_5$	[68]	**238**	Ganoleuconin H	$C_{30}H_{44}O_5$	[68]
239	Ganoleuconin I	$C_{30}H_{46}O_5$	[68]	**240**	Ganoleuconin J	$C_{38}H_{52}O_{13}$	[68]
241	Ganoleuconin K	$C_{36}H_{48}O_{12}$	[68]	**242**	Ganoleuconin L	$C_{36}H_{48}O_{11}$	[68]
243	Ganoleuconin M	$C_{51}H_{74}O_7$	[68]	**244**	Ganoleuconin N	$C_{51}H_{74}O_7$	[68]
245	Ganoleuconin O	$C_{51}H_{74}O_7$	[68]	**246**	Ganoleuconin P	$C_{51}H_{74}O_8$	[68]
247	Ganolucidic acid D	$C_{30}H_{44}O_6$	[69]	**248**	Ganolucidic acid E	$C_{30}H_{44}O_5$	[36]
249	Ganolucinin A	$C_{51}H_{72}O_8$	[70]	**250**	Ganolucinin B	$C_{51}H_{74}O_7$	[70]
251	Ganolucinin C	$C_{51}H_{74}O_7$	[70]	**252**	Ganorbiformin A	$C_{32}H_{48}O_8$	[71]
253	Ganorbiformin B	$C_{34}H_{50}O_7$	[71]	**254**	Ganorbiformin C	$C_{32}H_{48}O_6$	[71]
255	Ganorbiformin D	$C_{34}H_{50}O_8$	[71]	**256**	Ganorbiformin E	$C_{32}H_{48}O_6$	[71]
257	Ganorbiformin F	$C_{33}H_{50}O_6$	[71]	**258**	Ganorbiformin G	$C_{32}H_{46}O_5$	[71]
259	Ganosinensic acid A	$C_{27}H_{38}O_6$	[72]	**260**	Ganosinensic acid B	$C_{30}H_{42}O_7$	[72]
261	Ganosinensin A	$C_{51}H_{72}O_9$	[31]	**262**	Ganosinensin B	$C_{51}H_{72}O_9$	[31]
263	Ganosinensin C	$C_{51}H_{74}O_8$	[31]	**264**	Ganosporelactone A	$C_{30}H_{40}O_7$	[73]
265	Ganosporelactone B	$C_{30}H_{42}O_7$	[73]	**266**	Ganosporeric acid A	$C_{30}H_{38}O_8$	[74]
267	Ganotropic acid	$C_{30}H_{44}O_7$	[11]	**268**	Gibbosic acid A	$C_{30}H_{38}O_8$	[75]
269	Gibbosic acid B	$C_{30}H_{40}O_8$	[75]	**270**	Gibbosic acid C	$C_{30}H_{40}O_8$	[75]
271	Gibbosic acid D	$C_{30}H_{42}O_8$	[75]	**272**	Gibbosic acid E	$C_{30}H_{38}O_9$	[75]
273	Gibbosic acid F	$C_{31}H_{42}O_9$	[75]	**274**	Gibbosic acid G	$C_{30}H_{40}O_9$	[75]
275	Gibbosic acid H	$C_{30}H_{42}O_9$	[75]	**276**	Lanosta-7,9(11),24-trien-15-acetoxy-3-hydroxy-xo-26-oic acid	$C_{32}H_{46}O_6$	[7]
277	Lanosta-7,9(11),24-trien-3,15-diacetoxy-xo-26-oic acid	$C_{34}H_{48}O_7$	[7]	**278**	Lanosta-7,9(11),24-trien-3-acetox-y-15β-dihydroxy-26-oic acid	$C_{32}H_{48}O_6$	[7]
279	Lanosta-7,9(11),24-trien-3β,15β-triacetoxy-26-oic acid	$C_{36}H_{52}O_8$	[7]	**280**	Lanosta-7,9(11),24-trien-15-diacetoxy-3-hydroxy-26-oic acid	$C_{34}H_{50}O_7$	[23]
281	Lanosta-7,9(11),24-trien-3-acetoxy-15-hydroxy-xo-26-oic acid	$C_{32}H_{46}O_6$	[7]	**282**	Lanosta-7,9(11),24-trien-3,15-dihydroxy-26-oic acid	$C_{30}H_{46}O_4$	[23]
283	Leucocontextin A	$C_{30}H_{42}O_8$	[76]	**284**	Leucocontextin B	$C_{30}H_{44}O_8$	[76]
285	Leucocontextin C	$C_{30}H_{44}O_8$	[76]	**286**	Leucocontextin D	$C_{30}H_{42}O_8$	[76]
287	Leucocontextin E	$C_{30}H_{40}O_8$	[76]	**288**	Leucocontextin F	$C_{30}H_{44}O_7$	[76]
289	Leucocontextin G	$C_{30}H_{44}O_7$	[76]	**290**	Leucocontextin H	$C_{30}H_{44}O_7$	[76]
291	Leucocontextin I	$C_{30}H_{44}O_6$	[76]	**292**	Leucocontextin J	$C_{32}H_{44}O_9$	[76]
293	Leucocontextin K	$C_{32}H_{46}O_9$	[76]	**294**	Leucocontextin L	$C_{32}H_{44}O_9$	[76]
295	Leucocontextin M	$C_{32}H_{42}O_9$	[76]	**296**	Leucocontextin N	$C_{32}H_{46}O_8$	[76]
297	Leucocontextin O	$C_{32}H_{42}O_9$	[76]	**298**	Leucocontextin P	$C_{32}H_{46}O_9$	[76]
299	Leucocontextin Q	$C_{32}H_{46}O_9$	[76]	**300**	Leucocontextin R	$C_{30}H_{46}O_5$	[76]
301	Leucocontextin S	$C_{30}H_{48}O_6$	[77]	**302**	Leucocontextin T	$C_{30}H_{48}O_6$	[77]
303	Leucocontextin U	$C_{30}H_{50}O_6$	[77]	**304**	Leucocontextin V	$C_{30}H_{44}O_6$	[77]
305	Leucocontextin W	$C_{30}H_{48}O_5$	[77]	**306**	Leucocontextin X	$C_{36}H_{48}O_{11}$	[77]

(continued)

Table 3.1 (continued)

307	Lucialdehyde A	$C_{30}H_{46}O_2$	[78]	**308**	Lucialdehyde B	$C_{30}H_{44}O_3$	[78]
309	Lucialdehyde C	$C_{30}H_{46}O_3$	[78]	**310**	Lucidadiol	$C_{30}H_{48}O_3$	[29]
311	Lucidenic acid A	$C_{27}H_{38}O_6$	[79]	**312**	Lucidenic acid B	$C_{27}H_{38}O_7$	[79]
313	Lucidenic acid C	$C_{27}H_{40}O_7$	[79]	**314**	Lucidenic acid D	$C_{29}H_{38}O_8$	[38]
315	Lucidenic acid D1	$C_{27}H_{34}O_7$	[80]	**316**	Lucidenic acid E1	$C_{27}H_{38}O_7$	[80]
317	Lucidenic acid LM1	$C_{27}H_{40}O_6$	[38]	**318**	Lucidenic acid N	$C_{27}H_{40}O_6$	[81]
319	Lucidenic acid O	$C_{27}H_{38}O_7$	[82]	**320**	Lucidenic acid P	$C_{29}H_{42}O_8$	[83]
321	Lucidenic acid Q	$C_{32}H_{42}O_9$	[70]	**322**	Lucidenic acid R	$C_{32}H_{42}O_9$	[70]
323	Lucidenic acid S	$C_{30}H_{44}O_4$	[70]	**324**	Lucidenic lactone	$C_{27}H_{40}O_7$	[82]
325	Lucidone A	$C_{24}H_{34}O_5$	[80]	**326**	Lucidone B	$C_{24}H_{32}O_5$	[4]
327	Lucidone C	$C_{24}H_{36}O_5$	[44]	**328**	Lucidone D	$C_{24}H_{34}O_5$	[84]
329	Lucidumol A	$C_{30}H_{48}O_4$	[60]	**330**	Lucidumol B	$C_{30}H_{50}O_3$	[60]
331	Lucidumol C	$C_{30}H_{46}O_5$	[85]	**332**	Methyl 20(21)-dehydro-lucidenate A	$C_{28}H_{38}O_6$	[20]
333	Methyl 8,9-dihydro-ganoderate J	$C_{31}H_{46}O_7$	[12]	**334**	Methyl ganoderate D	$C_{31}H_{44}O_7$	[86]
335	Methyl ganoderate E	$C_{31}H_{42}O_7$	[86]	**336**	Methyl ganoderate F	$C_{33}H_{44}O_9$	[86]
337	Methyl ganoderate G	$C_{31}H_{46}O_8$	[80]	**338**	Methyl ganoderate H	$C_{33}H_{46}O_9$	[86]
339	Methyl ganoderate I	$C_{32}H_{46}O_9$	[34]	**340**	Methyl ganoderate J	$C_{31}H_{44}O_7$	[44]
341	Methyl ganoderate K2	$C_{31}H_{46}O_7$	[34]	**342**	Methyl ganoderate L	$C_{30}H_{46}O_8$	[69]
343	Methyl ganoderate M	$C_{30}H_{44}O_8$	[87]	**344**	Methyl ganoderate N	$C_{31}H_{44}O_8$	[87]
345	Methyl ganoderate O	$C_{31}H_{42}O_8$	[70]	**346**	Methyl ganoderate P	$C_{33}H_{46}O_9$	[70]
347	Methyl ganolucidate A	$C_{31}H_{46}O_6$	[80]	**348**	Methyl ganolucidate B	$C_{31}H_{48}O_6$	[80]
349	Methyl ganolucidate C	$C_{31}H_{48}O_7$	[88]	**350**	Methyl ganosinensate A	$C_{28}H_{40}O_6$	[72]
351	Methyl lucidenate E	$C_{30}H_{42}O_8$	[86]	**352**	Methyl lucidenate E2	$C_{30}H_{42}O_8$	[87]
353	Methyl lucidenate F	$C_{28}H_{38}O_6$	[86]	**354**	Methyl lucidenate G	$C_{28}H_{42}O_7$	[87]
355	Methyl lucidenate H	$C_{28}H_{42}O_7$	[87]	**356**	Methyl lucidenate I	$C_{28}H_{40}O_7$	[87]
357	Methyl lucidenate J	$C_{28}H_{40}O_8$	[87]	**358**	Methyl lucidenate K	$C_{28}H_{38}O_7$	[87]
359	Methyl lucidenate L	$C_{28}H_{40}O_7$	[87]	**360**	Methyl lucidenate M	$C_{28}H_{44}O_6$	[87]
361	Methyl lucidenate P	$C_{30}H_{44}O_8$	[87]	**362**	Methyl lucidenate Q	$C_{28}H_{42}O_6$	[87]
363	Petchinoids A	$C_{26}H_{36}O_6$	[89]	**364**	Petchinoids B	$C_{26}H_{38}O_6$	[89]
365	Petchinoids C	$C_{30}H_{46}O_5$	[89]	**366**	Sinensoic acid	$C_{30}H_{48}O_4$	[90]
367	Triacetate ganoderiol E	$C_{36}H_{54}O_7$	[36]	**368**	Tsugaric acid A	$C_{32}H_{50}O_4$	[91]
369	Tsugaric acid B	$C_{33}H_{52}O_5$	[50]	**370**	Tsugaric acid C	$C_{32}H_{50}O_5$	[92]
371	Tsugaric acid D	$C_{32}H_{48}O_5$	[48]	**372**	Tsugaric acid E	$C_{31}H_{46}O_4$	[48]
373	Tsugarioside B	$C_{37}H_{60}O_7$	[92]	**374**	Tsugarioside C	$C_{38}H_{58}O_8$	[92]

3.1.2.4 Supercritical Fluid Extraction

The technique of supercritical fluid is a new extraction method, in which supercritical carbon dioxide was used as solvent. Compared to the organic solvent extraction, the method of supercritical fluid has a lot of advantages. Supercritical carbon dioxide is the only solvent used in the extraction, so the process is organic solvent-free.

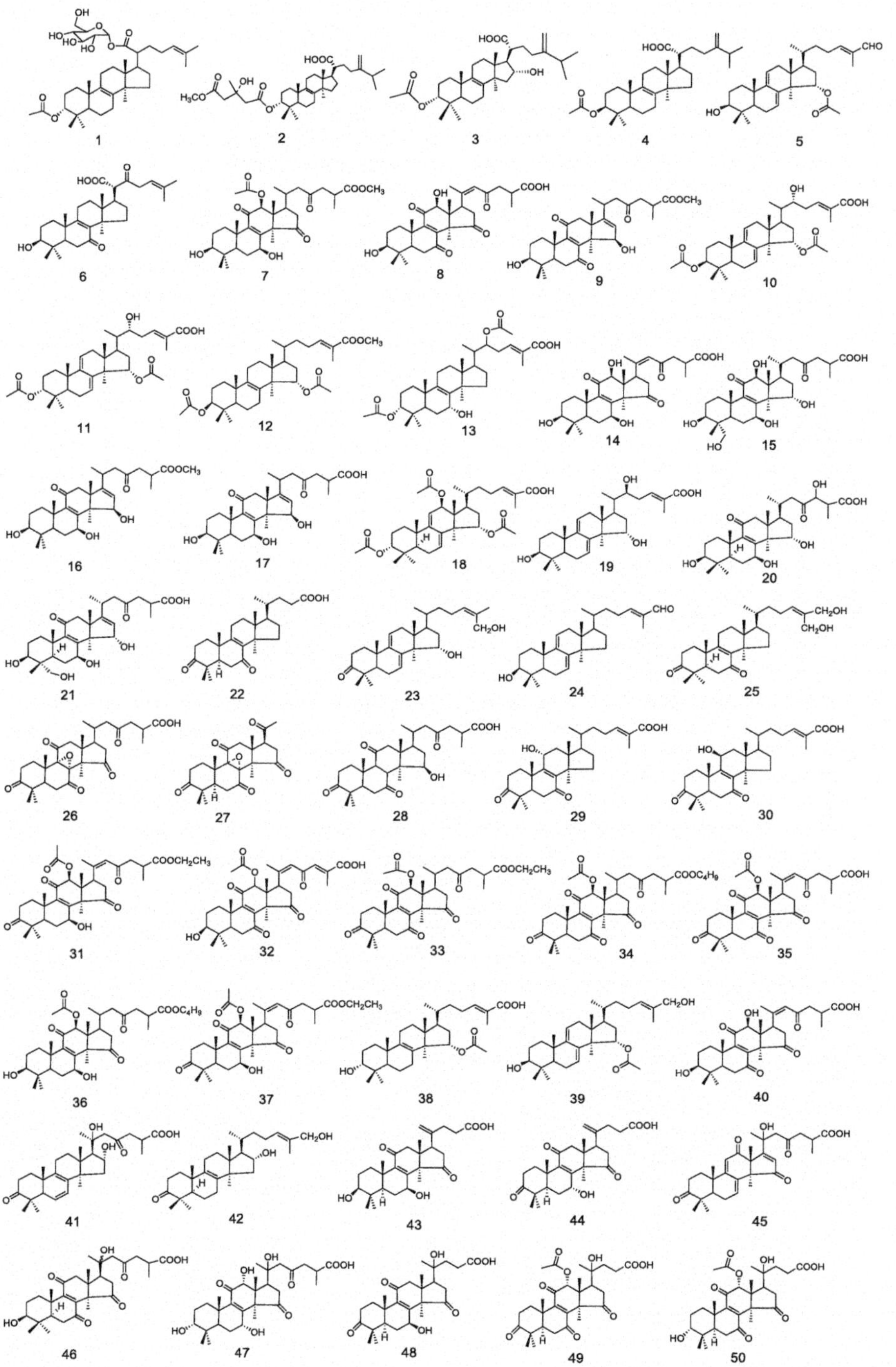

Fig. 3.1 Chemical structures of *Ganoderma* triterpenoids

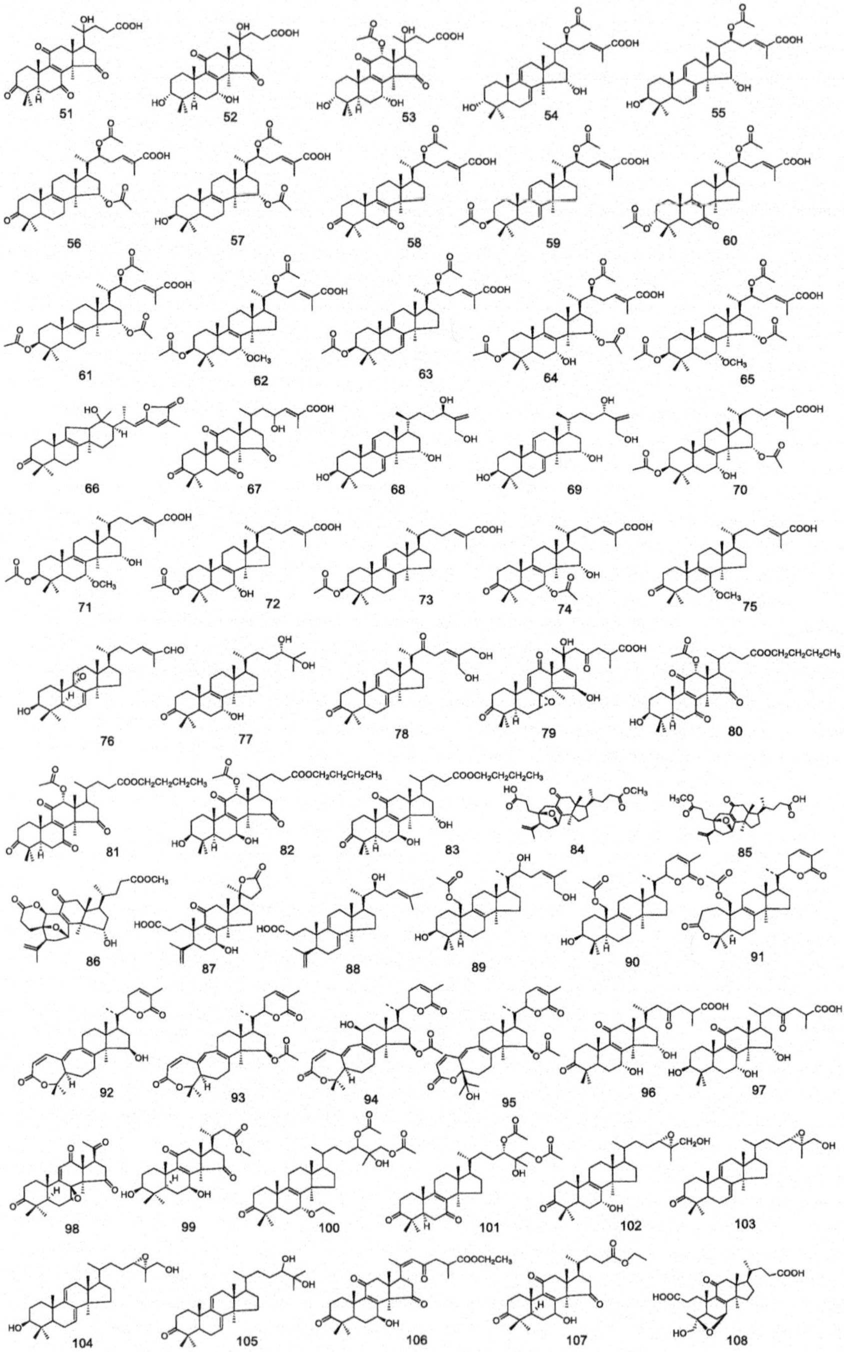

Fig. 3.1 (continued)

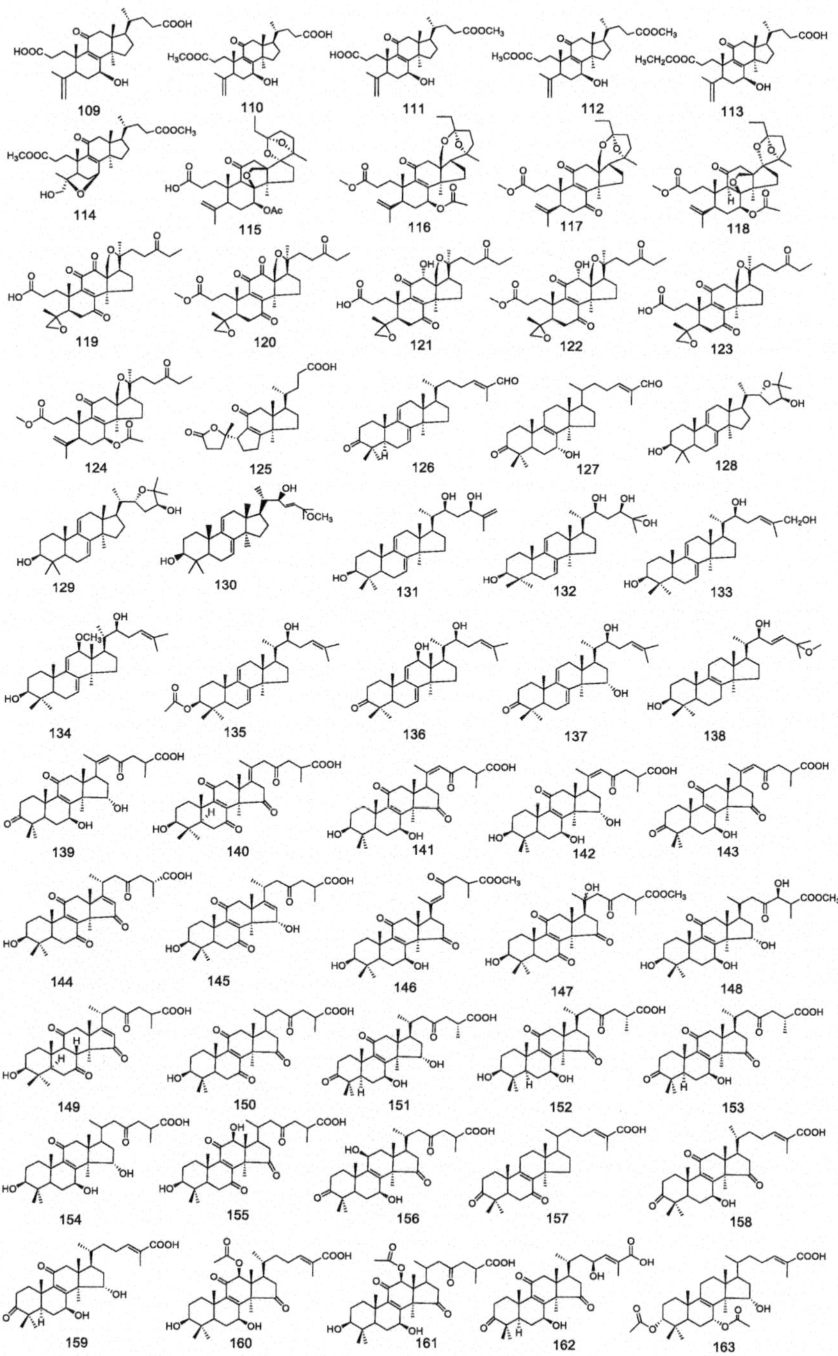

Fig. 3.1 (continued)

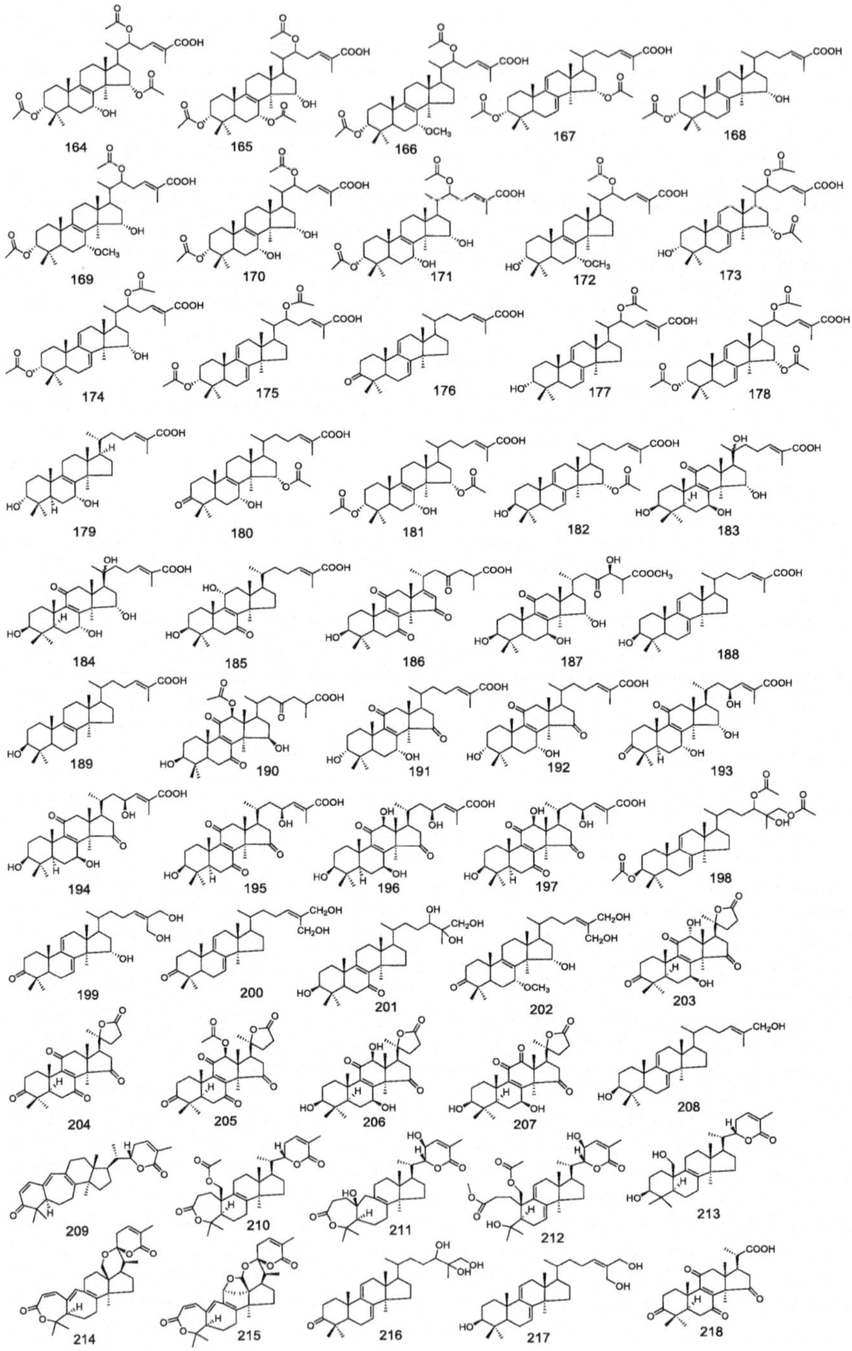

Fig. 3.1 (continued)

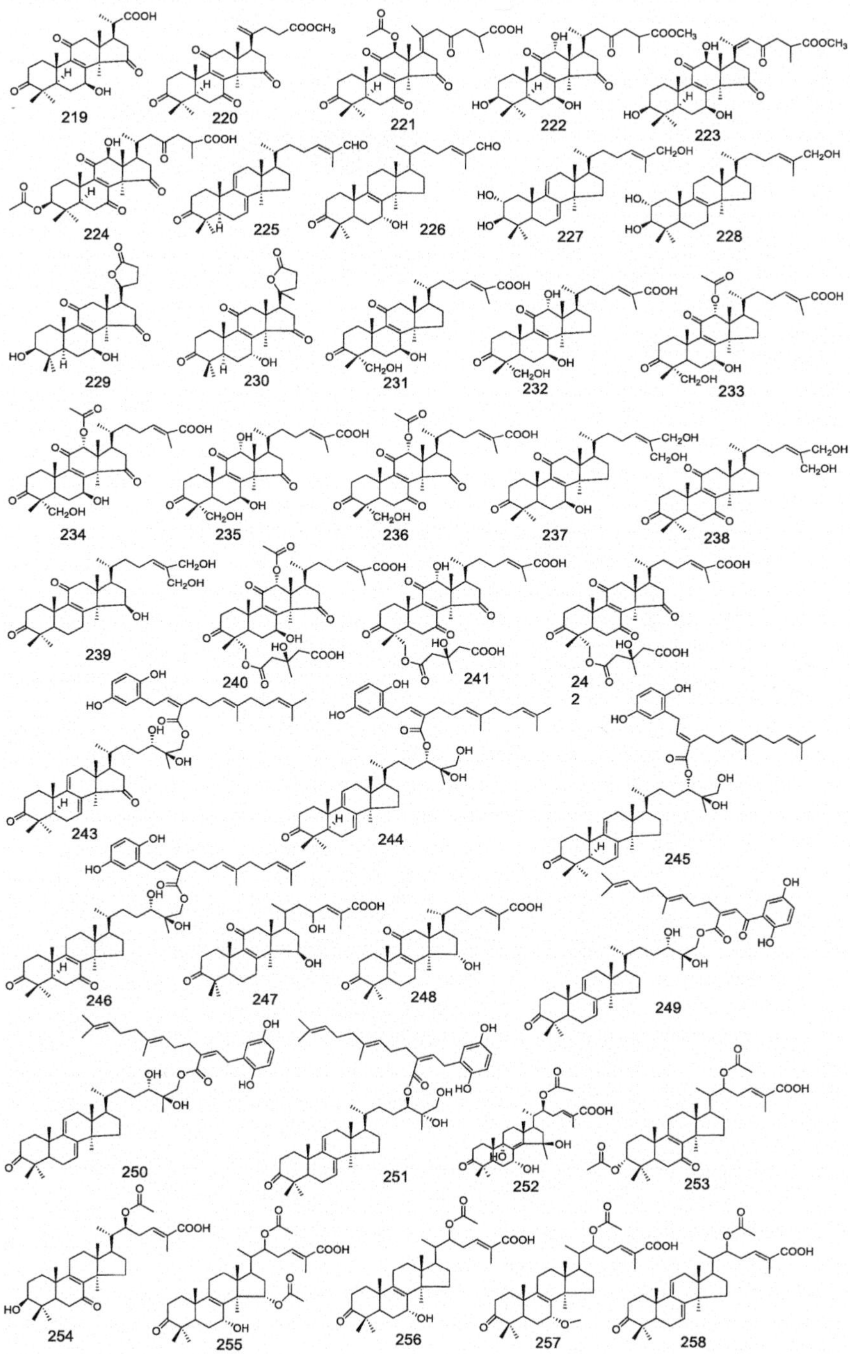

Fig. 3.1 (continued)

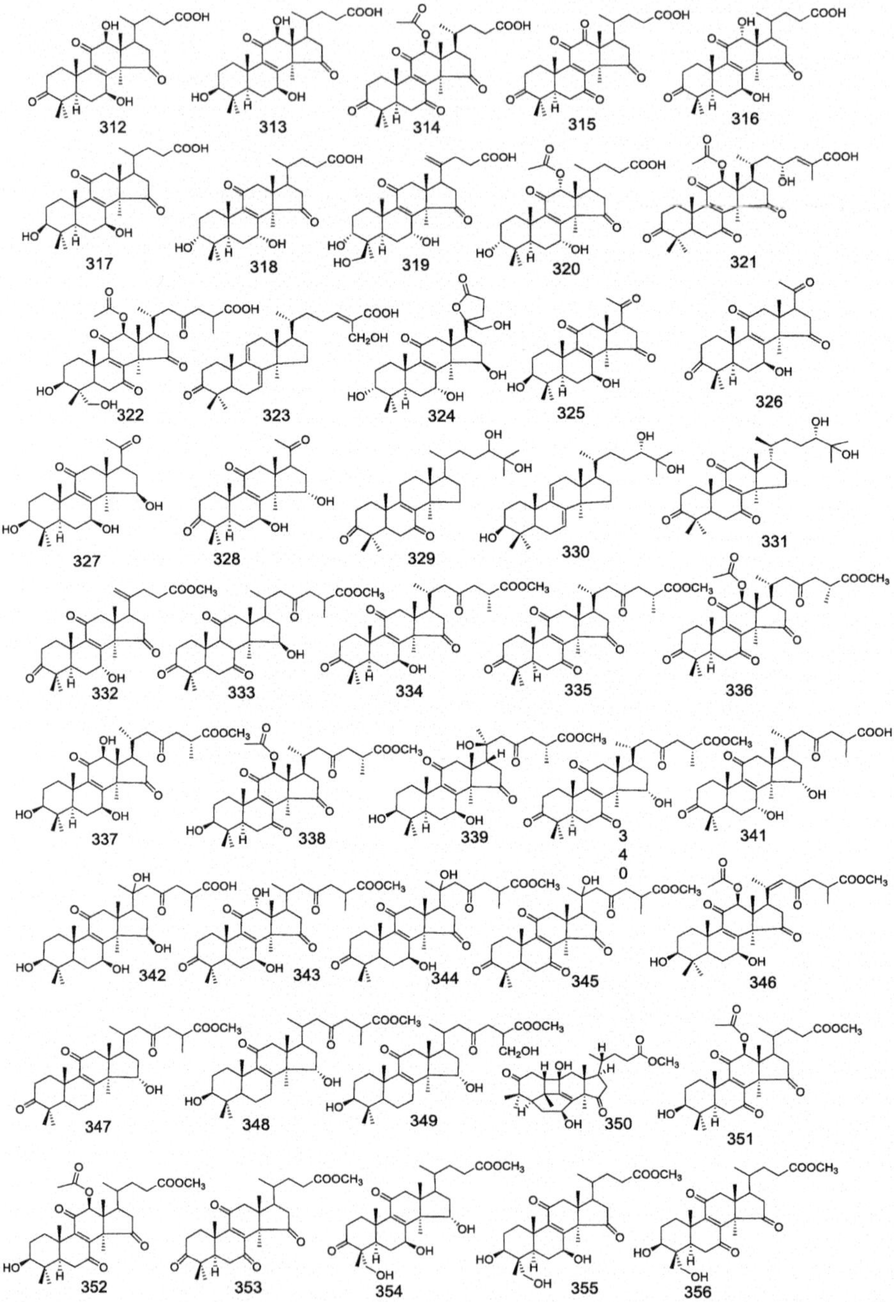

Fig. 3.1 (continued)

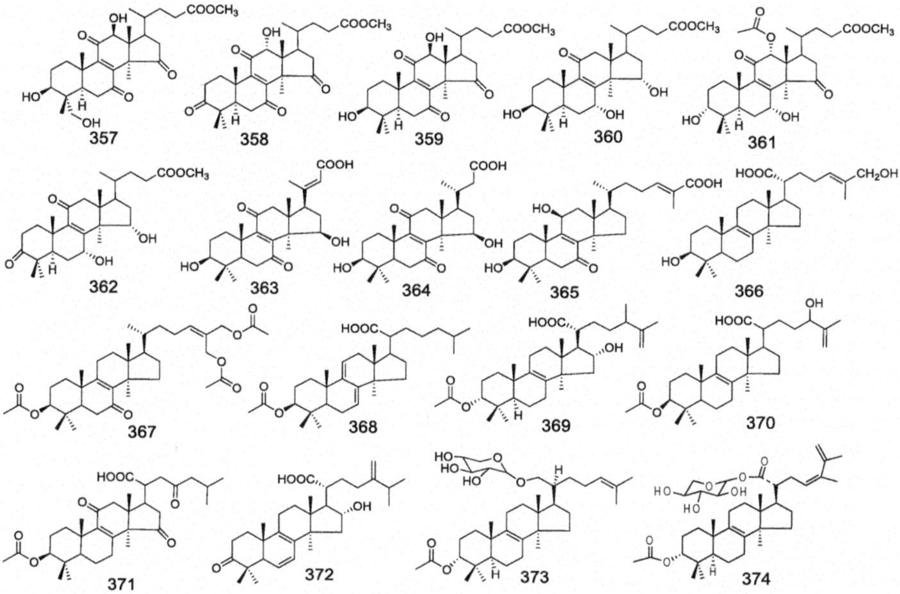

Fig. 3.1 (continued)

Due to the low extraction temperature (about 35 °C), the active ingredients are not easily changed in the process.

Zhang J et al. (2006) developed the method of supercritical fluid to extract the triterpenoids from *G. lucidum*. The extraction conditions are extraction pressure (15 MPa); extraction temperature (35 °C); extraction time (120 min); flow rate of CO_2 (1 mL/min), and the temperature of back pressure regulator (50 °C). They found that the structures and contents of triterpenoids extracted by supercritical carbon or methanol were quite similar, which demonstrated supercritical carbon dioxide could be used as a new green extraction solvent instead of organic solvent extraction [96].

3.2 Meroterpenoids

Meroterpenoids are hybrid natural products that originate from the shikimic acid and mevalonic acid biogenetical pathway. To our knowledge, it is not until 2000 that *Ganoderma* meroterpenoids (GMs), ganomycins A (**375**) and B (**376**), were reported from the mature fruiting bodies of *G. pfeifferi* [97]. Since then, with the development of new techniques, about 200 GMs were isolated from the genus *Ganoderma* (Tables 3.2, 3.3, 3.4, 3.5, 3.6, 3.7 and 3.8). GMs possessed diverse biological activities including inhibition of NO, antioxidant, anti-fibrotic, anti-allergic, anti-AChE, antimicrobial, cytotoxic activities, etc.

Table 3.2 Names, sources, and their bioactivities of GMs with 10- or 15-carbon side chains

No.	Names	Formulas	Bioactivities	Sources	References
375	Ganomycin A	$C_{21}H_{28}O_5$	Inhibition of several bacterial strains	*G. pfeifferi*	[97]
376	Ganomycin B	$C_{21}H_{28}O_4$	Inhibition of several bacterial strains; DPPH scavenging effect; inhibition of HMG-CoA reductase and α-glucosidase; inhibition of HMGs reductase, aldose reductase, sucrase, maltase, and α-glucosidase	*G. pfeifferi*	[97]
				G. capense	[98]
				G. leucocontextum	[99]
				G. lucidum	[100]
377	Fornicin D	$C_{16}H_{18}O_5$	Antioxidant activity	*G. cochlear*	[101, 102]
378	Cochlearin H	$C_{16}H_{18}O_5$	Antioxidant activity	*G. cochlear*	[103]
379	Chizhine D	$C_{16}H_{20}O_5$		*G. lucidum*	[104]
				G. cochlear	[102]
380	Cochlearin G	$C_{17}H_{22}O_5$	Antioxidant activity	*G. cochlear*	[103]
381	Applanatumol S	$C_{16}H_{20}O_6$		*G. applanatum*	[105]
				G. theaecolum	[106]
382	Applanatumol T	$C_{16}H_{18}O_7$		*G. applanatum*	[105]
383	Ganomycin F	$C_{21}H_{30}O_3$	DPPH scavenging effect	*G. capense*	[98]
384	Ganoleucin B	$C_{21}H_{28}O_4$		*G. leucocontextum*	[99]
385	Ganomycin J	$C_{21}H_{30}O_6$	Inhibition of HMGs reductase, aldose reductase, sucrase, maltase, and α-glucosidase	*G. lucidum*	[100]
386	Ganomycin E	$C_{21}H_{26}O_6$	DPPH radical scavenging activity	*G. capense* *G. australe*	[98] [107]
387	Ganomycin C	$C_{21}H_{26}O_5$	Antioxidant activity	*G. cochlear*	[101, 102]
				G. theaecolum	[106]
				G. australe	[107]
			Neuroprotection	*G. capense*	[98]
			DPPH scavenging effect	*G. leucocontextum*	[99]
			Inhibition of HMG-CoA reductase and α-glucosidase	*G. cochlear*	[108]
388	Cochlearin I	$C_{22}H_{28}O_5$	DPPH radical scavenging activity	*G. cochlear*	[103]
				G. australe	[107]
389	Cochlearol D	$C_{21}H_{26}O_5$		*G. cochlear*	[108]
390	Ganoresinain E	$C_{21}H_{28}O_7$		*G. resinaceum*	[109]
391	Ganotheaecolumol K	$C_{21}H_{26}O_6$	Inhibition of COX-2 and JAK3	*G. theaecolum*	[106]

(continued)

Table 3.2 (continued)

No.	Names	Formulas	Bioactivities	Sources	References
392	Isoganotheaecolumol I	$C_{21}H_{26}O_6$	Inhibition of COX-2 and JAK3	*G. theaecolum*	[106]
393	Ganotheaecolumol I	$C_{21}H_{26}O_6$	Inhibition of COX-2 and JAK3	*G. theaecolum*	[106]
394	Ganotheaecolumol J	$C_{21}H_{24}O_6$	Inhibition of COX-2 and JAK3	*G. theaecolum*	[106]
395	Ganocalidin D	$C_{21}H_{24}O_7$		*G. theaecolum*	[106]
				G. calidophilum	[110]
396	Cochlearol M	$C_{21}H_{26}O_6$		*G. cochlear*	[102]
397	Fornicin C	$C_{21}H_{28}O_5$		*G. cochlear*	[102]
				G. australe	[107]
			Moderate cytotoxic activity	*G. fornicatum*	[111]
			Inhibition of α-glucosidase	*G. leucocontextum*	[99]
398	Ganocalidin B	$C_{21}H_{26}O_7$		*G. calidophilum*	[110]
399	Ganocalidin F	$C_{21}H_{24}O_7$		*G. calidophilum*	[110]
400	Australeol D	$C_{21}H_{26}O_5$	Neuroprotection	*G. australe*	[107]
401	Cochlearol J	$C_{18}H_{24}O_5$		*G. cochlear*	[102]
402	Geranylhydroquinone	$C_{16}H_{22}O_2$		*G. cochlear*	[102]

3.2.1 *Chemical Structures of* Ganoderma *Meroterpenoids*

GM is composed of a 1,2,4-trisubstituted phenyl group and a mono- or sesquiterpene moiety (C10 or C15 chain or cyclic moiety). According to the characters of structures, GMs could be generally divided into seven types: (1) GMs with 10- or 15-carbon side chains, (2) GMs with lactone groups, (3) GMs with ether rings, (4) GMs with five- or six-membered carbon rings, (5) GMs with spiro rings, (6) GMs with bridged rings, and (7) dimeric GMs [135].

The diversity of the terpene moiety may be formed through oxidation, cyclization, isomerization, polymerization, etc. The ketone carbonyl at C-1′ and carboxyl or ester at C-10′ or C-14′ (e.g., **377**, **378**, **380**, **395**, **401**) were produced probably due to oxidation reaction.

An α, β-unsaturated γ-lactone connected to C-1′ of phenyl group might be formed through a nucleophilic addition reaction that happens between the carboxyl group at C-10′ or C-14′ and the ketone carbonyl at C-1′ (e.g., **403**, **406**). The ether rings were produced through a nucleophilic substitution reaction in which two hydroxyl groups were involved (e.g., **433**, **438**). The five-, six-, or seven-membered carbon rings were derived from a new C–C bond formation (**462**, **468**).

Table 3.3 Names, sources, and their bioactivities of GMs with lactone groups

No.	Names	Formulas	Bioactivities	Sources	References
403	(+)-Applanatumol U	$C_{16}H_{18}O_5$		*G. applanatum*	[105]
				G. theaecolum	[106]
404	(+)-Chizhine E	$C_{16}H_{16}O_5$		*G. lucidum*	[104]
				G. theaecolum	[106]
405	Fornicin A	$C_{16}H_{18}O_4$	Moderate cytotoxic activity	*G. fornicatum*	[111]
				G. sinense	[112]
406	(+)-Zizhine A	$C_{21}H_{26}O_5$		*G. sinense*	[112]
				G. resinaceum	[109]
				G. theaecolum	[106]
				G. australe	[107]
407	Ganoresinain B	$C_{21}H_{26}O_5$		*G. resinaceum*	[109]
				G. australe	[107]
408	Ganocalidin C	$C_{21}H_{24}O_6$		*G. calidophilum*	[110]
409	Ganocalidin E	$C_{21}H_{26}O_6$		*G. calidophilum*	[110]
410	Ganomycin I	$C_{21}H_{26}O_5$	Inhibition of MCP-1 expression; inhibition of HMGs reductase, aldose reductase, sucrase, maltase, and α-glucosidase	*G. lucidum*	[104]
			Anti-proliferation of neural stem cells	*G. lucidum*	[100]
				G. leucocontextum	[99]
				G. lingzhi	[113]
411	(+)-Ganoleucin C	$C_{21}H_{24}O_5$	Inhibition of α-glucosidase	*G. leucocontextum*	[99]
				G. theaecolum	[106]
412	(+)-Chizhine F	$C_{21}H_{24}O_5$	Inhibition of MCP-1 expression	*G. lucidum*	[104]
413	(+)-Zizhine B	$C_{23}H_{28}O_7$		*G. sinense*	[112]
414	(+)-Zizhine C	$C_{23}H_{30}O_7$		*G. sinense*	[112]
415	(+)-Zizhine D	$C_{23}H_{28}O_7$		*G. sinense*	[112]
416	(+)-Zizhine E	$C_{23}H_{30}O_7$		*G. sinense*	[112]
417	(+)-Zizhine F	$C_{24}H_{30}O_8$		*G. sinense*	[112]
418	Fornicin B	$C_{22}H_{28}O_5$	Moderate cytotoxic activity	*G. fornicatum*	[111]
				G. cochlear	[102]
			Inhibition of MCP-1 expression	*G. lucidum*	[104]
			DPPH scavenging effect	*G. capense*	[98]
			Inhibition of HMG-CoA reductase and α-glucosidase	*G. leucocontextum*	[99]
419	(+)-Fornicin E	$C_{22}H_{28}O_6$	DPPH scavenging effect	*G. capense*	[98]
			Renoprotective effects	*G. australe*	[107]

(continued)

Table 3.3 (continued)

No.	Names	Formulas	Bioactivities	Sources	References
420	Ganotheaecolumol G	$C_{22}H_{26}O_7$	Inhibition of COX-2 and JAK3	*G. theaecolum*	[106]
421	Ganotheaecolumol H	$C_{22}H_{26}O_6$		*G. theaecolum*	[106]
422	Ganotheaecolumol E	$C_{21}H_{24}O_6$		*G. theaecolum*	[106]
423	Ganotheaecolumol F	$C_{21}H_{26}O_5$		*G. theaecolum*	[106]
424	Cochlearol L	$C_{17}H_{20}O_6$		*G. cochlear*	[102]
425	Austraeol C	$C_{22}H_{28}O_6$		*G. australe*	[107]
426	(+)-Lucidulactone B	$C_{17}H_{18}O_6$		*G. lucidum*	[114]
				G. applanatum	[115]
427	Austraeol E	$C_{23}H_{30}O_6$		*G. australe*	[107]
428	Austraeol F	$C_{23}H_{30}O_6$		*G. australe*	[107]
429	Chizhine A	$C_{16}H_{18}O_5$		*G. lucidum*	[104]
430	Chizhine B	$C_{16}H_{18}O_5$		*G. lucidum*	[104]
431	Chizhine C	$C_{16}H_{18}O_6$		*G. lucidum*	[104]
432	Ganoleucin A	$C_{21}H_{28}O_5$	Inhibition of α-glucosidase	*G. leucocontextum*	[99]

3.2.1.1 Extraction and Identification of *Ganoderma* Meroterpenoids

Ganoderma fungi were usually extracted with 95% EtOH; then the extract was partitioned between water and petroleum ether, $CHCl_3$ and EtOAc. Thus, the GMs were enriched in $CHCl_3$ and EtOAc fraction, which was further separated by column chromatography on silica gel, C-18, MCI, and Sephadex LH-20. Then the compounds were finally purified using semi-preparative HPLC system (Figs. 3.2, 3.3, 3.4, 3.5, 3.6, 3.7 and 3.8).

3.3 Steroids

Steroids are a kind of compounds that commonly occur in *Ganoderma* fungi, in which ergosterols, stigmasterols, sitosterol, and daucosterol were found. Among them, ergosterols are the most common compounds in the genus *Ganoderma*. There are more than 30 different ergosterols isolated from it (Table 3.9, Fig. 3.9).

Table 3.4 Names, sources, and their bioactivities of GMs with ether rings

No.	Names	Formulas	Bioactivities	Sources	References
433	(+)-Cochlearin B	$C_{21}H_{28}O_4$	Antioxidant activity	*G. cochlear*	[103]
434	(±)-Cochlearin D	$C_{21}H_{28}O_3$	Antioxidant activity	*G. cochlear*	[103]
435	(+)-Lingzhine A		Promotion of proliferation of neural stem cells	*G. lingzhi*	[113]
436	(+)-Lingzhine E	$C_{16}H_{18}O_6$	Promotion of proliferation of neural stem cells	*G. lingzhi* *G. australe*	[113] [107]
437	(+)-Lingzhine F	$C_{16}H_{18}O_6$	Promotion of proliferation of neural stem cells	*G. austral* *G. lingzhi*	[107] [113]
438	(+)-Applanatumol P	$C_{17}H_{20}O_7$		*G. applanatum*	[105]
439	(+)-Applanatumol Q	$C_{17}H_{22}O_7$		*G. applanatum*	[105]
440	(+)-Applanatumol R	$C_{16}H_{20}O_7$		*G. applanatum*	[105]
441	(±)-Ganocapensin A	$C_{21}H_{24}O_6$	DPPH scavenging effect	*G. capense*	[98]
442	Ganoresinain A	$C_{21}H_{24}O_5$	Neuroprotection	*G. resinaceum* *G. australe*	[109] [107]
443	Ganocapensin B	$C_{21}H_{28}O_5$	DPPH scavenging effect	*G. capense*	[98]
444	Australeol A	$C_{21}H_{26}O_5$	Neuroprotection	*G. australe*	[107]
445	Australeol B	$C_{21}H_{26}O_5$	Neuroprotection	*G. australe*	[107]
446	Ganocalidin A	$C_{21}H_{24}O_6$	Anti-allergic activity	*G. calidophilum*	[110]
447	Petchiether A	$C_{21}H_{28}O_5$	Inhibition of fibronectin expression	*G. australe* *G. petchii*	[107] [116]
448	Petchiether B	$C_{21}H_{28}O_5$	Inhibition of fibronectin expression	*G. petchii*	[116]
449	Cochlearin C	$C_{21}H_{26}O_4$	Antioxidant activity	*G. cochlear*	[103]
450	(±)-Cochlearin E	$C_{21}H_{28}O_3$	Antioxidant activity Protection against renal fibrosis	*G. cochlear*	[103] [102]
451	Cochlearin F	$C_{15}H_{12}O_4$	Antioxidant activity	*G. cochlear*	[103]
452	Ganotheaecolumol C	$C_{20}H_{26}O_5$	Inhibition of COX-2 and JAK3	*G. theaecolum*	[106]
453	Ganotheaecolumol D	$C_{20}H_{26}O_5$	Inhibition of COX-2 and JAK3	*G. theaecolum*	[106]
454	Cochlearol I	$C_{20}H_{26}O_4$		*G. cochlear*	[102]
455	Applanatumol Z1	$C_{13}H_{12}O_5$		*G. applanatum*	[105]
456	Cochlearol C	$C_{20}H_{24}O_4$		*G. cochlear*	[108]
457	Ganotheaecolumol A	$C_{21}H_{26}O_6$		*G. theaecolum* *G. australe*	[106] [107]
458	Ganotheaecolumol B	$C_{21}H_{26}O_6$		*G. theaecolum* *G. australe*	[106] [107]
459	Cochlearol E	$C_{21}H_{26}O_7$		*G. cochlear*	[102]
460	Cochlearol F	$C_{21}H_{26}O_7$		*G. cochlear*	[102]
461	Cochlearol K	$C_{16}H_{18}O_5$	Protection against renal fibrosis	*G. cochlear*	[102]

Table 3.5 Names, sources, and their bioactivities of GMs with five-, six-, or seven membered carbon rings

No.	Names	Formulas	Bioactivities	Sources	References
462	Applanatumol V	$C_{16}H_{16}O_6$		*G. applanatum*	[105]
463	Applanatumol W	$C_{17}H_{18}O_6$		*G. applanatum*	[105]
464	Applanatumol X	$C_{13}H_{12}O_5$		*G. applanatum*	[105]
465	Applanatumol Y	$C_{14}H_{14}O_5$		*G. applanatum*	[105]
466	Applanatumol Z	$C_{14}H_{16}O_6$		*G. applanatum*	[105]
467	Applanatumol Z2	$C_{14}H_{12}O_5$		*G. applanatum*	[105]
468	Applanatumol K	$C_{16}H_{18}O_7$		*G. applanatum*	[105]
469	Applanatumol L	$C_{17}H_{20}O_7$		*G. applanatum*	[105]
470	Applanatumol M	$C_{16}H_{16}O_6$		*G. applanatum*	[105]
471	Applanatumol N	$C_{16}H_{18}O_7$		*G. applanatum* *G. australe* *G. theaecolum*	[105] [107] [117]
472	Applanatumol O	$C_{16}H_{16}O_6$		*G. applanatum*	[105]
473	Chizhiol A	$C_{16}H_{18}O_6$		*G. lucidum*	[118]
474	Ganotheaecoloid L	$C_{16}H_{18}O_6$		*G. theaecolum*	[117]
475	(+)-Ganotheaecoloid M	$C_{16}H_{20}O_7$		*G. theaecolum*	[117]
476	(−)-Ganotheaecoloid N	$C_{17}H_{20}O_6$		*G. theaecolum*	[117]
477	Petchiene A	$C_{16}H_{18}O_6$		*G. petchii* *G. resinaceum* *G. theaecolum*	[119] [109] [117]
478	Petchiene B	$C_{15}H_{16}O_4$	Augmentation of intracellular free calcium concentration	*G. petchii*	[119]
479	Petchiene C	$C_{15}H_{16}O_3$		*G. petchii*	[119]
480	Petchiene E	$C_{15}H_{12}O_5$		*G. petchii*	[119]
481	Lingzhine C	$C_{15}H_{12}O_4$	Antiproliferation of neural stem cells	*G. lingzhi* *G. resinaceum*	[113] [109]

(continued)

Table 3.5 (continued)

No.	Names	Formulas	Bioactivities	Sources	References
482	(±)-Lingzhine B	$C_{15}H_{14}O_4$	Promotion of proliferation of neural stem cells	*G. lingzhi*	[113]
483	(−)-Ganotheaecoloid A	$C_{20}H_{26}O_5$		*G. theaecolum*	[117]
484	(−)-Ganotheaecoloid B	$C_{20}H_{26}O_5$		*G. theaecolum*	[117]
485	Cochlearol G	$C_{21}H_{26}O_6$		*G. cochlear*	[102]
486	Cochlearol H	$C_{21}H_{26}O_6$		*G. cochlear*	[102]
487	Ganotheaecoloid C	$C_{21}H_{28}O_6$		*G. theaecolum* *G. australe*	[117] [107]
488	Ganotheaecoloid D	$C_{21}H_{26}O_6$		*G. theaecolum* *G. australe*	[117] [107]
489	Ganotheaecoloid E	$C_{21}H_{28}O_6$		*G. theaecolum*	[117]
490	(−)-Ganotheaecoloid F	$C_{21}H_{26}O_6$		*G. theaecolum*	[117]
491	Ganotheaecoloid G	$C_{21}H_{26}O_5$	Neuroprotection	*G. theaecolum* *G. australe*	[117] [107]
492	Ganotheaecoloid H	$C_{21}H_{26}O_5$		*G. theaecolum*	[117]
493	Ganoresinain C	$C_{21}H_{26}O_5$		*G. resinaceum*	[109]
494	Ganotheaecoloid I	$C_{21}H_{26}O_6$		*G. theaecolum*	[117]
495	(+)-Ganotheaecoloid J	$C_{22}H_{28}O_6$	Inhibition of COX-2	*G. theaecolum* *G. australe*	[117] [107]
496	Ganotheaecoloid K	$C_{21}H_{28}O_5$		*G. theaecolum*	[117]
497	(+)-Cochlearin A	$C_{21}H_{26}O_3$	Antioxidant activity	*G. cochlear*	[103]
498	Ganoresinain D	$C_{21}H_{26}O_6$		*G. resinaceum*	[109]

3.4 Alkaloids and Nucleosides

In 1990, ganoine (**608**) and ganodine (**609**) were firstly isolated from *G. capense*. Up to now, no more than 30 alkaloids were obtained, including monocyclic and polycyclic alkaloids, purine, pyrimidine, and cerebrosides. Among them, (±)-sinensilactam A (**629**), a rare hybrid metabolite possessing activity of inhibiting Smad3 phosphorylation, was isolated from the fruiting bodies of *G. sinensis* (Table 3.10; Fig. 3.10).

Table 3.6 Names, sources, and their bioactivities of GMs with spiro rings

No.	Names	Formulas	Bioactivities	Sources	References
499	Spiroapplanatumine A	$C_{16}H_{14}O_7$		*G. applanatum*	[120]
500	Spiroapplanatumine C	$C_{17}H_{16}O_7$		*G. applanatum*	[120]
501	Spiroapplanatumine E	$C_{17}H_{16}O_7$		*G. applanatum*	[120]
502	Spiroapplanatumine G	$C_{16}H_{14}O_6$	Inhibition of JAK3	*G. applanatum*	[120]
503	Spiroapplanatumine I	$C_{17}H_{16}O_6$		*G. applanatum*	[120]
504	Spiroapplanatumine B	$C_{16}H_{14}O_7$		*G. applanatum*	[120]
505	Spiroapplanatumine D	$C_{17}H_{16}O_7$		*G. applanatum*	[120]
506	Spiroapplanatumine F	$C_{17}H_{16}O_7$		*G. applanatum*	[120]
507	Spiroapplanatumine H	$C_{16}H_{14}O_6$	Inhibition of JAK3	*G. applanatum*	[120]
508	Spiroapplanatumine J	$C_{17}H_{18}O_7$		*G. applanatum*	[120]
509	Spiroapplanatumine K	$C_{17}H_{18}O_6$	Inhibition of aldose reductase	*G. applanatum* *G. leucocontextum*	[120] [121]
510	Spiroapplanatumine L	$C_{16}H_{16}O_6$	Inhibition of aldose reductase	*G. applanatum* *G. leucocontextum*	[120] [121]
511	Spiroapplanatumine M	$C_{16}H_{16}O_6$		*G. applanatum*	[120]
512	(+)-Spiroapplanatumine N	$C_{16}H_{14}O_6$		*G. applanatum*	[120]
513	Spiroapplanatumine O	$C_{17}H_{16}O_6$		*G. applanatum*	[120]
514	(−)-Spiroapplanatumine N	$C_{16}H_{14}O_6$		*G. applanatum*	[120]
515	Spiroapplanatumine P	$C_{17}H_{20}O_6$		*G. applanatum*	[120]
516	Spiroapplanatumine Q	$C_{14}H_{14}O_6$		*G. applanatum*	[120]
517	(+)-Spirolingzhine A	$C_{16}H_{18}O_6$	Promotion of proliferation of neural stem cells; inhibition of aldose reductase and HMG-CoA reductase	*G. lingzhi* *G. applanatum* *G. leucocontextum*	[113] [120] [121]
518	(+)-Spirolingzhine B	$C_{16}H_{18}O_6$	Promotion of proliferation of neural stem cells	*G. lingzhi* *G. applanatum*	[113] [120]
519	(+)-Spirolingzhine C	$C_{15}H_{14}O_6$	Promotion of proliferation of neural stem cells	*G. lingzhi*	[113]
520	Spirolingzhine D	$C_{16}H_{16}O_6$	Promotion of proliferation of neural stem cells; inhibition of aldose reductase	*G. lingzhi* *G. applanatum* *G. leucocontextum*	[113] [120] [121]
521	Petchiene D	$C_{15}H_{16}O_4$	Augmentation of intracellular free calcium concentration	*G. petchii*	[119]

Table 3.7 Names, sources, and their bioactivities of GMs with bridged rings

No.	Names	Formulas	Bioactivities	Sources	References
522	(±)-Ganoderin A	$C_{17}H_{20}O_7$	Antioxidant activity	*G. cochlear*	[101]
523	Applanatumol H	$C_{16}H_{18}O_7$		*G. applanatum*	[105]
524	Applanatumol I	$C_{16}H_{16}O_8$		*G. applanatum*	[105]
525	Applanatumol J	$C_{15}H_{15}ClO_6$		*G. applanatum*	[105]
526	Applanatumol D	$C_{17}H_{16}O_8$		*G. applanatum*	[105]
527	Applanatumol E	$C_{18}H_{22}O_8$		*G. applanatum*	[105]
528	Applanatumol G	$C_{20}H_{26}O_8$		*G. applanatum*	[105]
529	Applanatumol F	$C_{18}H_{22}O_8$		*G. applanatum*	[105]
530	Lingzhilactone A	$C_{18}H_{20}O_7$		*G. lingzhi*	[122]
531	Lingzhilactone B	$C_{16}H_{16}O_7$	Renoprotective effect	*G. lingzhi* *G. applanatum*	[122] [115]
532	Lingzhilactone C	$C_{20}H_{26}O_8$		*G. lingzhi*	[122]
533	Applanatumol Z3	$C_{19}H_{22}O_9$		*G. applanatum*	[115]
534	Applanatumol Z4	$C_{19}H_{22}O_9$		*G. applanatum*	[115]
535	(±)-Cochlactone A	$C_{21}H_{26}O_5$	Anti-inflammatory activity	*G. cochlear*	[123]
536	(±)-Cochlactone B	$C_{21}H_{26}O_5$	Anti-inflammatory activity	*G. cochlear*	[123]
537	Applanatumol C	$C_{15}H_{14}O_6$	Inhibition of COX-2	*G. applanatum*	[105]
538	(−)-Lingzhiol	$C_{15}H_{14}O_6$	Renoprotective effect	*G. lucidum* *G. australe* *G. applanatum*	[124] [107] [115]
539	(±)-Ganocochlearin A	$C_{20}H_{24}O_3$	Antioxidant activity	*G. cochlear*	[101]
540	(±)-Ganocochlearin B	$C_{15}H_{16}O_3$	Antioxidant activity	*G. cochlear*	[101]
541	(±)-Ganocochlearin C	$C_{21}H_{22}O_3$	Antioxidant activity	*G. cochlear*	[101]
542	(±)-Ganocochlearin D	$C_{21}H_{22}O_4$	Antioxidant activity	*G. cochlear*	[101]
543	Lingzhine D	$C_{16}H_{14}O_4$	Promotion of proliferation of neural stem cells	*G. lingzhi*	[113]
544	(±)-Ganocin A	$C_{21}H_{24}O_4$		*G. cochlear*	[125]
545	(±)-Ganocin B	$C_{20}H_{22}O_3$		*G. cochlear*	[125]
546	(±)-Ganocin C	$C_{19}H_{20}O_3$		*G. cochlear*	[125]

(continued)

Table 3.7 (continued)

No.	Names	Formulas	Bioactivities	Sources	References
547	(±)-Ganocin D	$C_{20}H_{22}O_3$	Anti-AChE activity	*G. cochlear*	[125]
548	Cochlearol A	$C_{15}H_{14}O_7$		*G. cochlear*	[126]
549	Cochlearol B	$C_{22}H_{28}O_3$	Renoprotective effect	*G. cochlear*	[126]
550	Applanatumol A	$C_{16}H_{16}O_6$	Anti-renal fibrosis	*G. applanatum*	[127]
551	Applanatumol B	$C_{16}H_{18}O_6$	Anti-renal fibrosis	*G. applanatum*	[127]

Table 3.8 Names, sources, and bioactivities of dimeric GMs

No.	Names	Formulas	Bioactivities	Sources	References
552	(+)-Ganoapplanin	$C_{24}H_{20}O_{10}$	Inhibition on intracellular Ca^{2+} via T-type voltage-gated calcium channels	*G. applanatum*	[128]
553	Applanatumin A	$C_{32}H_{30}O_{12}$	Antifibrotic activity	*G. applanatum*	[129]
554	Cochlearoid A	$C_{40}H_{48}O_8$	Inhibition on intracellular Ca^{2+} via T-type voltage-gated calcium channels	*G. cochlear*	[130]
555	Cochlearoid B	$C_{38}H_{46}O_6$		*G. cochlear*	[130]
556	Cochlearoid C	$C_{43}H_{54}O_7$	Inhibition on intracellular Ca^{2+} via T-type voltage-gated calcium channels	*G. cochlear*	[130]
557	Cochlearoid D	$C_{43}H_{54}O_7$		*G. cochlear*	[130]
558	Cochlearoid E	$C_{40}H_{48}O_8$		*G. cochlear*	[130]
559	Ganodilactone	$C_{42}H_{50}O_7$	Inhibition of COX-2 and pancreatic lipase; cytotoxicity	*G. leucocontextum* *G. cochlear*	[131] [132]
560	Spirocochlealactone A	$C_{42}H_{50}O_8$	Inhibition of COX-2; cytotoxicity	*G. cochlear*	[132]
561	Spirocochlealactone B	$C_{37}H_{42}O_8$	Inhibition of COX-2	*G. cochlear*	[132]
562	Spirocochlealactone C	$C_{37}H_{42}O_8$	Inhibition of COX-2	*G. cochlear*	[132]
563	Cochlearoid F	$C_{42}H_{52}O_7$	Renoprotective effect	*G. cochlear*	[13]
564	Cochlearoid G	$C_{42}H_{52}O_7$	Renoprotective effect	*G. cochlear*	[133]
565	Cochlearoid H	$C_{28}H_{30}O_7$	Renoprotective effect	*G. cochlear*	[133]
566	Cochlearoid I	$C_{28}H_{32}O_6$	Renoprotective effect	*G. cochlear*	[133]
567	Cochlearoid J	$C_{28}H_{32}O_6$		*G. cochlear*	[133]
568	Cochlearoid K	$C_{23}H_{22}O_7$	Renoprotective effect	*G. cochlear*	[133]
569	Gancochlearol A	$C_{42}H_{54}O_7$	Inhibition of COX-2; cytotoxicity	*G. cochlear*	[134]
570	Gancochlearol B	$C_{37}H_{46}O_7$	Inhibition of COX-2; cytotoxicity	*G. cochlear*	[134]
571	Ganoleucin D	$C_{32}H_{30}O_{11}$	Inhibition of aldose reductase and HMG-CoA reductase	*G. leucocontextum*	[121]

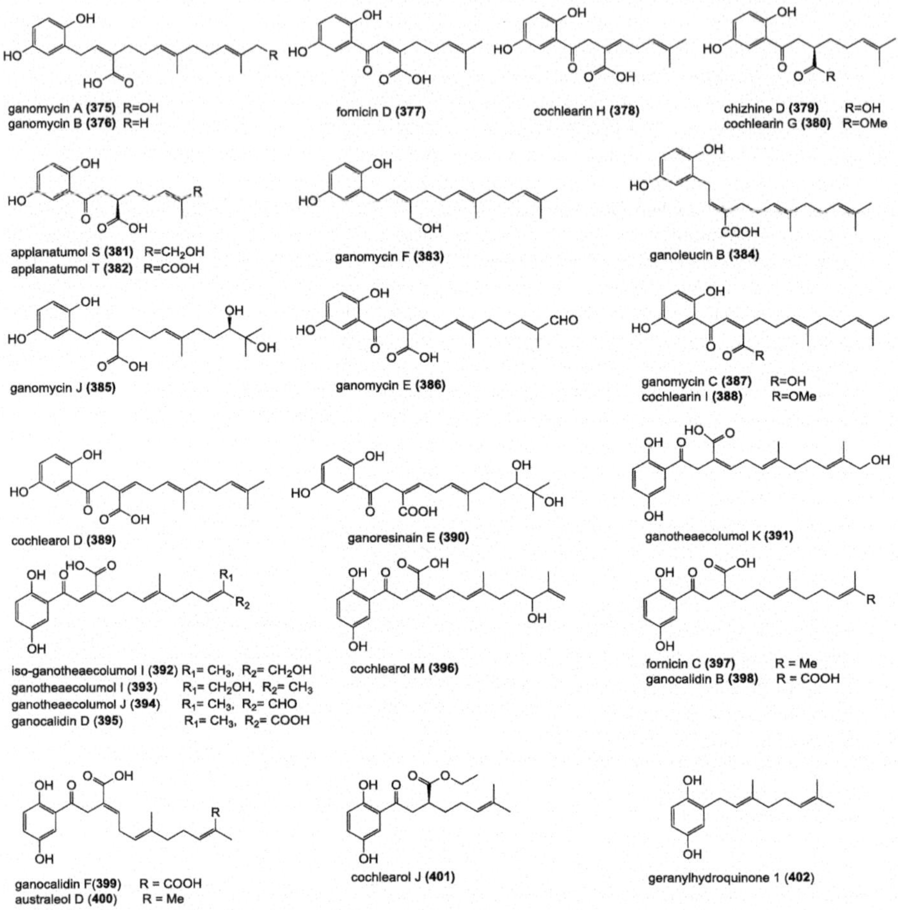

Fig. 3.2 Chemical structures of *Ganoderma* meroterpenoids with a 10-carbon or 15-carbon chain

3.5 Polysaccharides of *Ganoderma lucidum*

Polysaccharides are polymeric carbohydrate molecules composed of long chains of at least ten monosaccharide units bound together by glycosidic linkages.

The relative molecular weights of polysaccharides and glycoprotein are 10^3–10^6 Da. The monosaccharide units are composed of homoglycans or heteroglycans mainly containing D-glucose, and the linkages between them are usually $1 \rightarrow 3$, $1 \rightarrow 4$, or $1 \rightarrow 6$ [160]. Some polysaccharides contained glycoprotein. The sugar chains are connected with serine or threonine in peptide chain through *O*-glycosidic bonds in glycoproteins. In addition, the methods of isolation and identification of polysaccharides from *G. lucidum* were introduced here. Polysaccharides of *G. lucidum* show a lot of bioactivities, including immunoregulation, inhibition of vas-

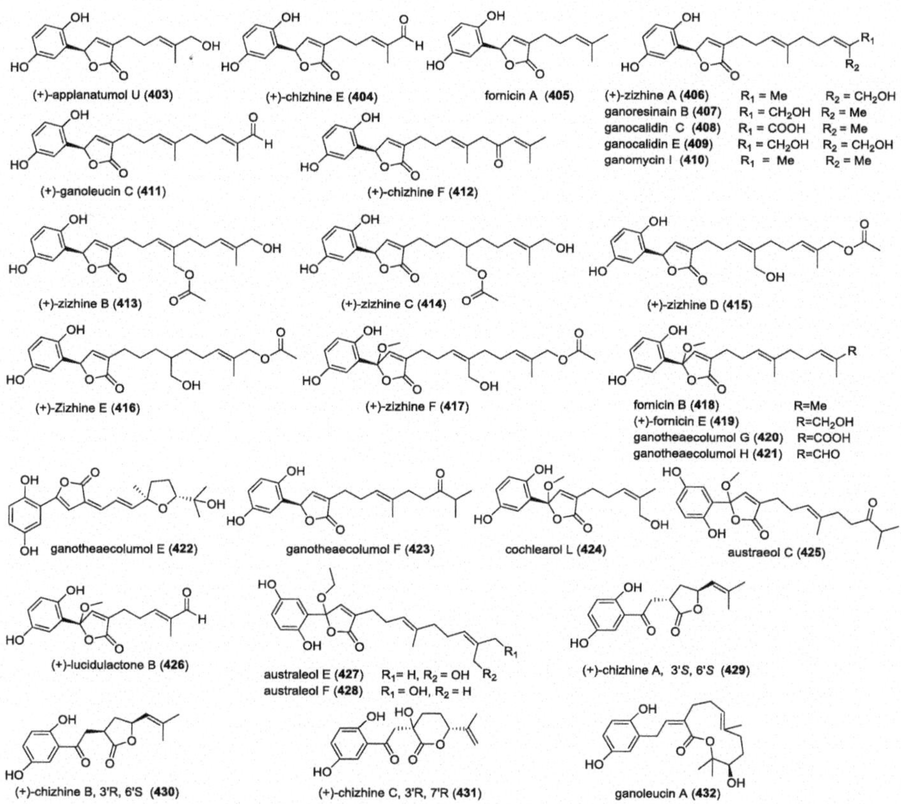

Fig. 3.3 Chemical structures of *Ganoderma* meroterpenoids with a lactone

cular proliferation, promotion of insulin release, hypoglycemic effect, anti-cancer, anti-oxidation, anti-aging, etc. [160, 161].

3.5.1 The Chemical Characterization of Polysaccharides Isolated from G. lucidum

There are more than 200 polysaccharides isolated and purified from the fruiting bodies, spores, mycelia, and cultivation broth of *G. lucidum*. Most polysaccharides are composed of homoglucans (α/β-glucans), glycoproteins, and heteropolysaccharides with glucose, galactose, fucose, mannose, and arabinose combined in different proportions and types of glycosidic linkages.

The structural characterization of polysaccharides isolated from *G. lucidum* is shown in Table 3.11.

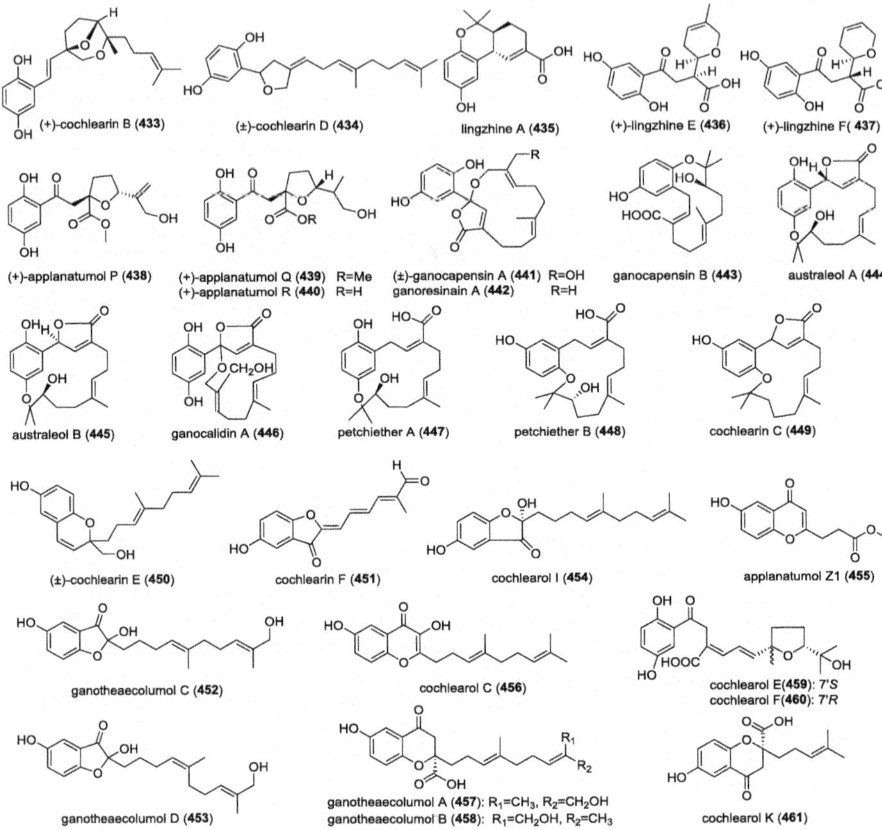

Fig. 3.4 Chemical structures of *Ganoderma* meroterpenoids with an ether ring

3.5.2 The Methods of Isolation of Polysaccharides of G. lucidum

The methods of separation of polysaccharides include fractional precipitation, ion exchange chromatography, ultrafiltration, etc.

3.5.2.1 Fractional Precipitation

According to the different solubility of polysaccharides in different concentration of MeOH and EtOH, the different molecules of polysaccharides could be precipitated, successively.

First, the total polysaccharides water solution (1~3%) was centrifuged to remove impurities. Then, EtOH was dropped into the water solution [1: 0.5(V/V)]. The solution was kept overnight and then centrifuged and precipitated by 33% EtOH. The

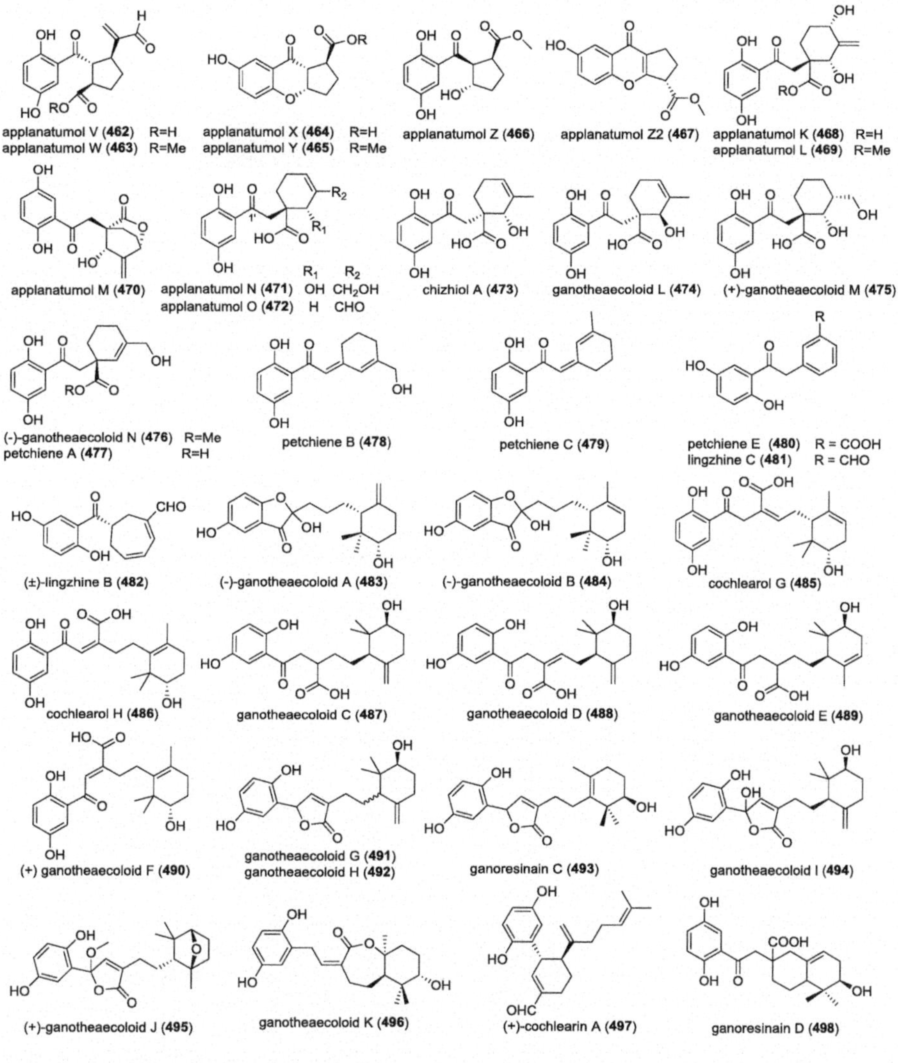

Fig. 3.5 Chemical structures of *Ganoderma* meroterpenoids with a five-membered or six-membered carbon ring

precipitation was dried to get the first fraction. Then EtOH was added into supernatant [1:1 (V/V)], and the same steps were followed to get the second fraction. At last, EtOH was added into supernatant [1:1.5 (V/V)], and the third fraction was obtained in the same method [174, 175].

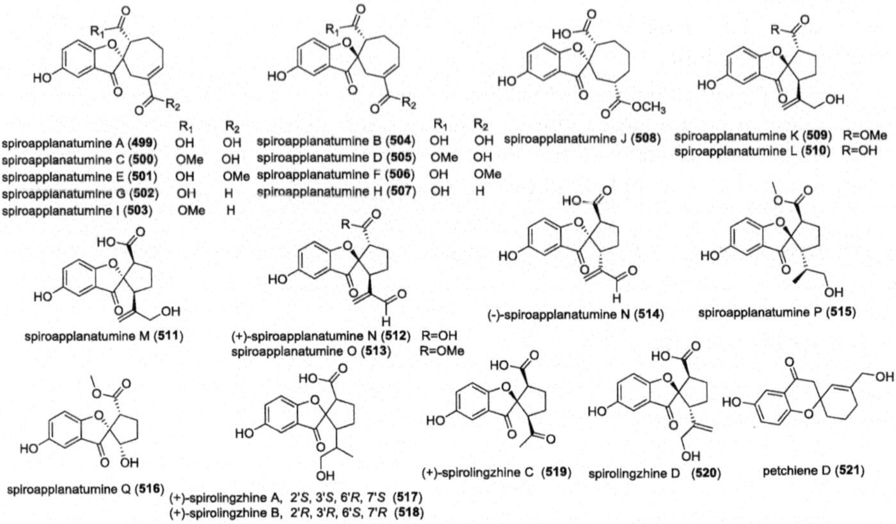

Fig. 3.6 Chemical structures of *Ganoderma* meroterpenoids with spiro ring

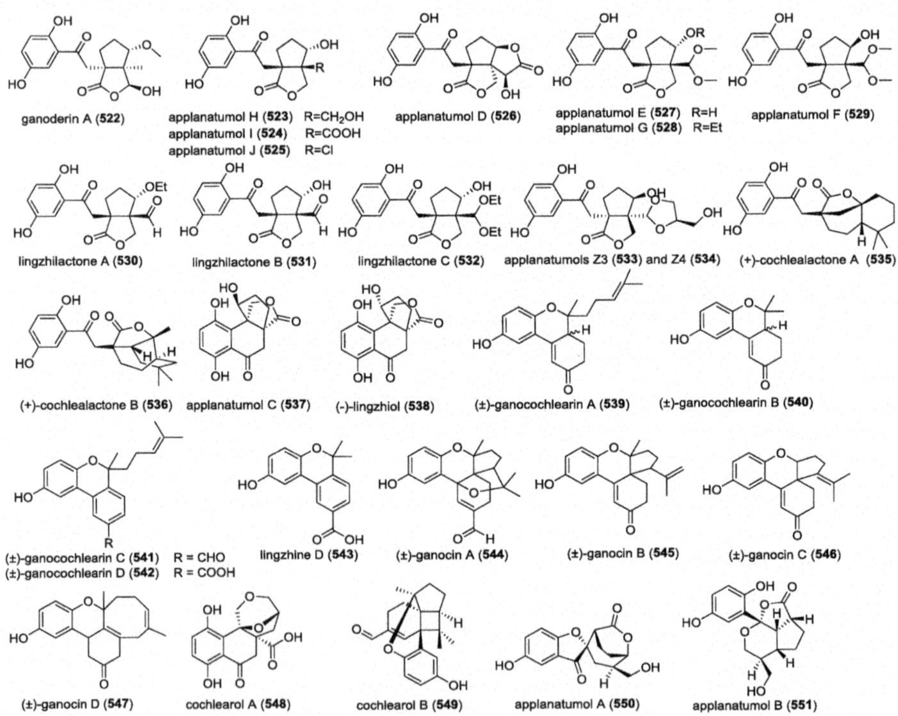

Fig. 3.7 Chemical structures of *Ganoderma* meroterpenoids with bridge ring

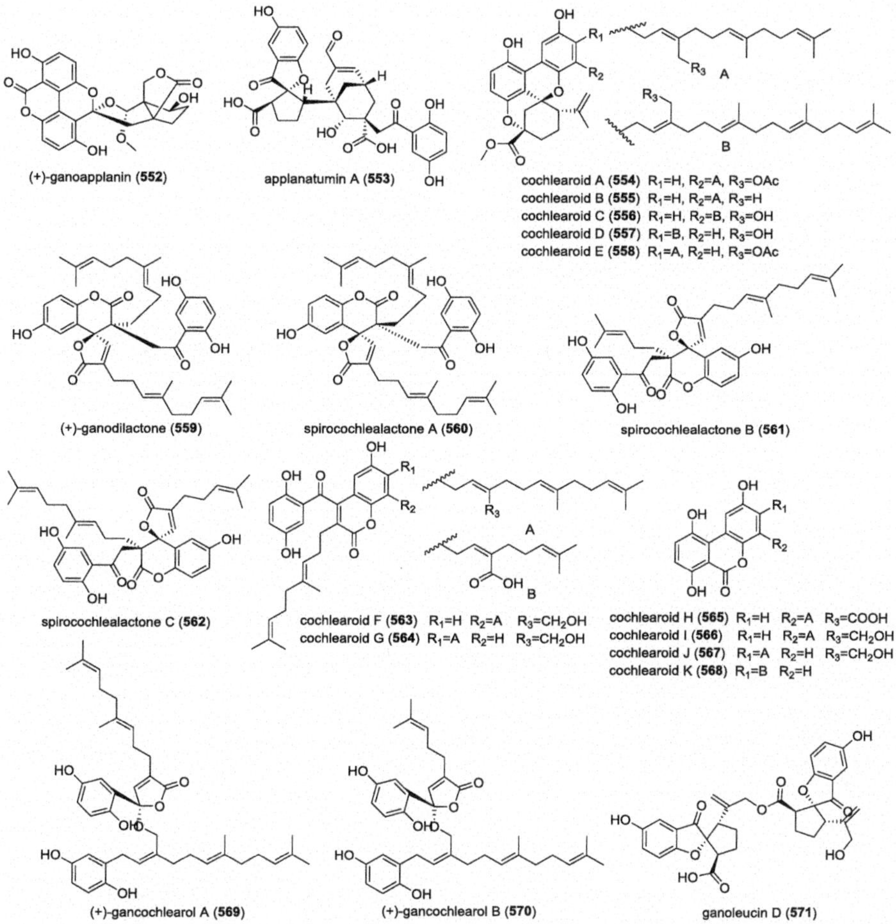

Fig. 3.8 Chemical structures of dimeric *Ganoderma* meroterpenoids

3.5.2.2 Ion-Exchange Chromatography

There are three most popular ion exchangers: cellulose ion-exchanger, Sephadex ion-exchanger and Sepharose ion-exchanger [176]. The latter two are called anion exchanger because the negative ions in samples could be exchanged when they are connected with the positive ions of diethylaminoethyl (DEAE). They are all wildly used in the separation of polysaccharides [177, 178].

Table 3.9 The names and formulas of steroids

No.	Names	Formulas	References
572	Stigmasta-7,22-diene-3*β*,5*α*,6*α*-triol	$C_{28}H_{46}O_3$	[136, 137]
573	(22*E*,24*R*)-ergosta-7,22-diene-3*β*,5*α*,6*β*-triol	$C_{28}H_{46}O_3$	[136–138]
574	22*E*-6*β*-methoxyergosta-7,22-diene-3*β*,5*α*-diol	$C_{29}H_{48}O_3$	[136, 138]
575	Ergosta-7,22-dien-3*β*,4*α*-diol	$C_{28}H_{46}O_2$	[138]
576	(3*β*,5*α*,6*β*,22*E*)-ergosta-7,22-diene-3,5,6-triol 6-stearate	$C_{46}H_{80}O_4$	[138]
577	Ergosta-7,22-dien-3*β*-ol	$C_{28}H_{46}O$	[139, 144]
578	3*β*,5*α*-dihydroxy-(22*E*,24*R*)-ergosta-7,22-dien-6-one	$C_{28}H_{44}O_3$	[145]
579	Ergosterol	$C_{28}H_{44}O$	[137, 138, 144]
580	Ergosta-7,22-dien-3*β*-yl palmitate	$C_{44}H_{74}O_2$	[137, 139, 141]
581	Ergosta-7,22-dien-3*β*-yl linoleate	$C_{46}H_{78}O_2$	[143]
582	3*β*,5*α*,9*α*-trihydroxy-(22*E*,24*R*)-ergosta-7,22-dien-6-one	$C_{28}H_{44}O_4$	[138]
583	Ergosta-7,22-dien-3*β*,5*α*, 6*β*, 9*α*-tetraol	$C_{28}H_{46}O_4$	[145]
584	Ergosta-7,22-dien-2*β*,4*α*-diol	$C_{28}H_{46}O_2$	[145]
585	Ergosta-7,22-dien-2*β*,3*α*,9*α*-triol	$C_{28}H_{46}O_3$	[139–143]
586	22*E*-7*α*-methoxy-5*α*,6*α*-epoxyergosta-8(14),22-dien-3*β*-ol	$C_{29}H_{46}O_3$	[136]
587	(5*α*,6*β*,15*β*,22*E*)-6-ethoxy-5,15-dihydroxyergosta-7,22-dien-3-one	$C_{30}H_{48}O_4$	[146]
588	Ergosta-4,6,8(14),22-tetraen-3-one	$C_{28}H_{40}O$	[136, 137] [139, 140]
589	Ganodermaside D	$C_{28}H_{40}O_2$	[139, 146]
590	Cyathisterol	$C_{28}H_{42}O_2$	[136]
591	Ergosta-7,22-dien-3-one	$C_{28}H_{44}O$	[136, 139–141]
592	(14*β*,22*E*)-9,14-dihydroxyergosta-4,7,22-triene-3,6-dione	$C_{28}H_{40}O_4$	[146]
593	Calvasterol B	$C_{28}H_{40}O_4$	[146]
594	Ergosta-4,7,22-triene-3,6-dione	$C_{28}H_{40}O_2$	[147]
595	Calvasterol A	$C_{28}H_{38}O_3$	[146]
596	Ergosta-6,22-dien-3*β*,5*α*,8*α*-triol	$C_{28}H_{46}O_3$	[138]
597	(22*E*,24*R*)-6*β*-methoxyergosta-7,9(11),22-triene-3*β*,5*α*-diol	$C_{29}H_{46}O_3$	[136]
598	(22*E*,24*R*)-ergosta-7,9(11),22-triene-3*β*,5*α*,6*α*-triol	$C_{28}H_{44}O_3$	[136, 137]
599	(22*E*,24*R*)-ergosta-7,9(11),22-triene-3*β*,5*α*,6*β*-triol	$C_{28}H_{44}O_3$	[136, 138]
600	(22*E*,24*S*)-5*α*,8*α*-epidioxy-24-methyl-cholesta-6,22-dien-3*β*-ol	$C_{28}H_{44}O_3$	[136, 142]
601	5*α*,8*α*-epidioxyergosta-6,22-dien-3*β*-yl linoleate	$C_{46}H_{74}O_4$	[140, 141] [143, 144]
602	(22*E*,24*S*)-5*α*,8*α*-epidioxy-24-methyl-cholesta-6,9(11),22-trien-3*β*-ol	$C_{28}H_{42}O_3$	[136, 142, 143]
603	5*α*,9*α*-epidioxyergosta-6,8(14),22-triene-3*β*-ol	$C_{28}H_{42}O_3$	[148]
604	6,9-epidioxyergosta-7,22-dien-3*β*-ol	$C_{28}H_{44}O_2$	[149]
605	Tetraoxycitricolic acid	$C_{21}H_{32}O_6$	[136]
606	11*α*-hydroxy-21-hydroxy-demethylincisterol A3	$C_{21}H_{32}O_6$	[148]
607	Demethylincisterol A3	$C_{21}H_{32}O_4$	[149]

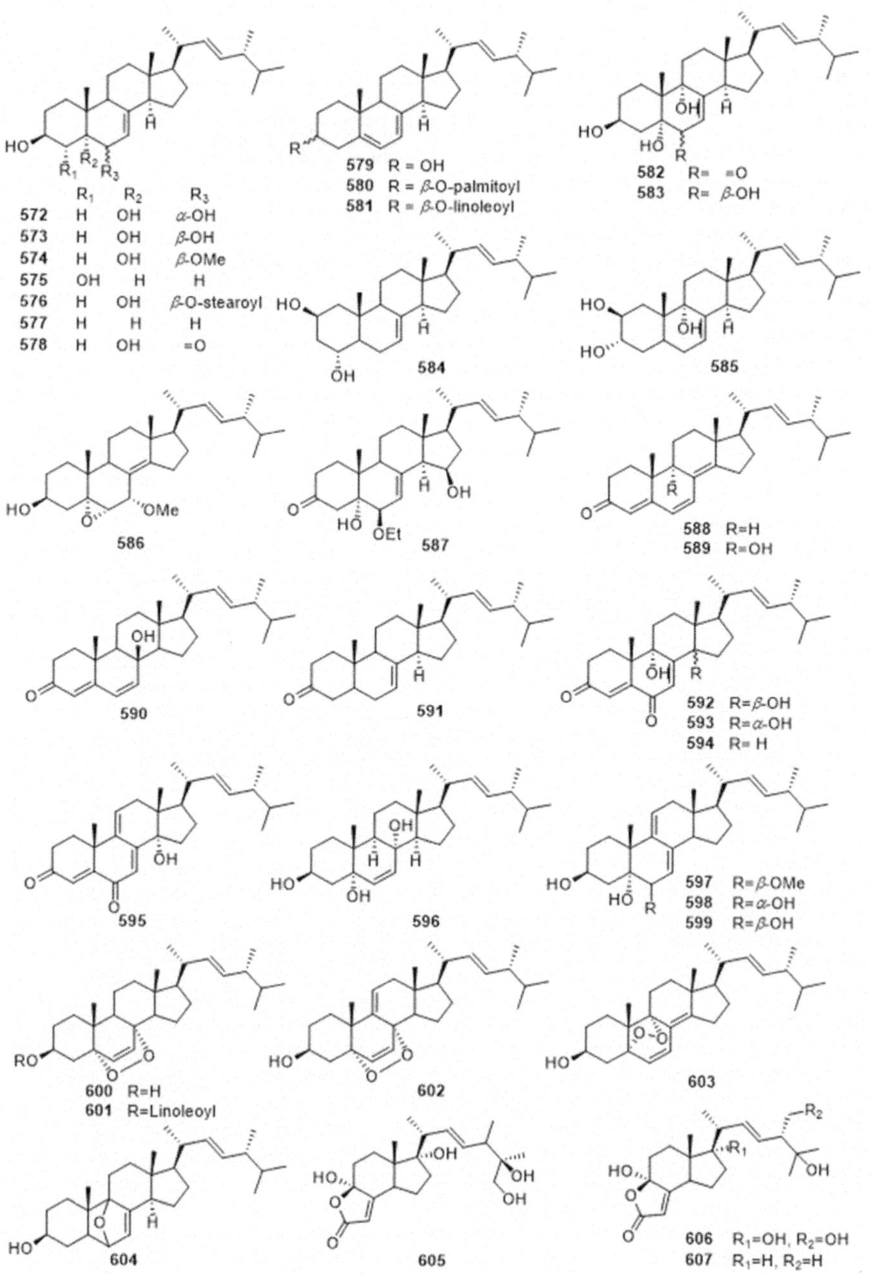

Fig. 3.9 Chemical structures of *Ganoderma* steroids

Table 3.10 Names, formulas, and sources of alkaloids

No.	Names	Formulas	Sources	References
608	Ganoine	$C_{11}H_{17}O_2N$	*G. capense*	[150]
609	Ganodine	$C_{14}H_{15}O_2N$	*G. capense*	[150]
610	Ganocochlearine A	$C_{14}H_{13}O_2N$	*G. cochlear*	[151]
611	Ganocochlearine B	$C_{15}H_{15}O_2N$	*G. cochlear*	[151]
612	Ganocochlearine C	$C_{19}H_{19}O_4N$	*G. cochlear*	[152]
613	Ganocochlearine D	$C_{17}H_{17}O_4N$	*G. cochlear*	[152]
614	Ganocochlearine E	$C_{17}H_{17}O_3N$	*G. cochlear*	[152]
615	Ganocochlearine F	$C_{15}H_{11}O_4N$	*G. cochlear*	[152]
616	Ganocochlearine G	$C_{15}H_{15}O_3N$	*G. cochlear*	[152]
617	Ganocochlearine H	$C_{15}H_{13}O_2N$	*G. cochlear*	[152]
618	Ganocochlearine I	$C_{15}H_{13}O_3N$	*G. cochlear*	[152]
619	Ganoapplanatumine B	$C_{16}H_{15}O_4N$	*G. cochlear*	[152]
			G. applanatum	[153]
620	Lucidimine A (ganocalicine A)	$C_{16}H_{15}O_3N$	*G. lucidum* *G. calidophilum*	[154] [110]
621	Lucidimine B (ganocalicine A)	$C_{15}H_{13}O_2N$	*G. lucidum* *G. calidophilum*	[154] [110]
622	Lucidimine C	$C_{16}H_{15}O_3N$	*G. cochlear* *G. lucidum*	[152] [154]
623	Lucidimine D	$C_{17}H_{17}O_4N$	*G. lucidum*	[154]
624	Sinensine	$C_{15}H_{15}O_3N$	*G. sinense*	[155, 156]
625	Sinensine B	$C_{14}H_{13}O_2N$	*G. sinense*	[155]
626	Sinensine C	$C_{14}H_{13}O_3N$	*G. sinense*	[155]
627	Sinensine D	$C_{14}H_{11}O_3N$	*G. sinense*	[155]
628	Sinensine E	$C_{15}H_{13}O_3N$	*G. cochlear* *G. sinense*	[152] [155]
629	(±)-Sinensilactam A	$C_{20}H_{21}O_8N$	*G. sinense*	[157]
630	Ganoderpurine	$C_{11}H_{15}ON_5$	*G. capense*	[150]
631	Uracil	$C_4H_4N_2O_2$	*G. lucidum*	[158]
632	Uridine	$C_9H_{12}N_2O_6$	*G. lucidum*	[158]
633	Adenine	$C_5H_5N_5$	*G. lucidum*	[158]
634	Adenosine	$C_{10}H_{13}N_5O_4$	*G. lucidum*	[158]
635	(4*E*,8*E*)-N-D-2′-hydroxypalmitoyl-1-*O*-*β*-D-glucopyranosyl-9-methyl-4,8-sphingadienine	$C_{41}H_{77}O_9N$	*G. lucidum*	[159]
636	(4*E*,8*E*)-N-D-2′-hydroxystearoyl-1-*O*-*β*-D-glucopyranosyl-9-methyl-4,8-sphingadienine	$C_{43}H_{81}O_9N$	*G. lucidum*	[159]

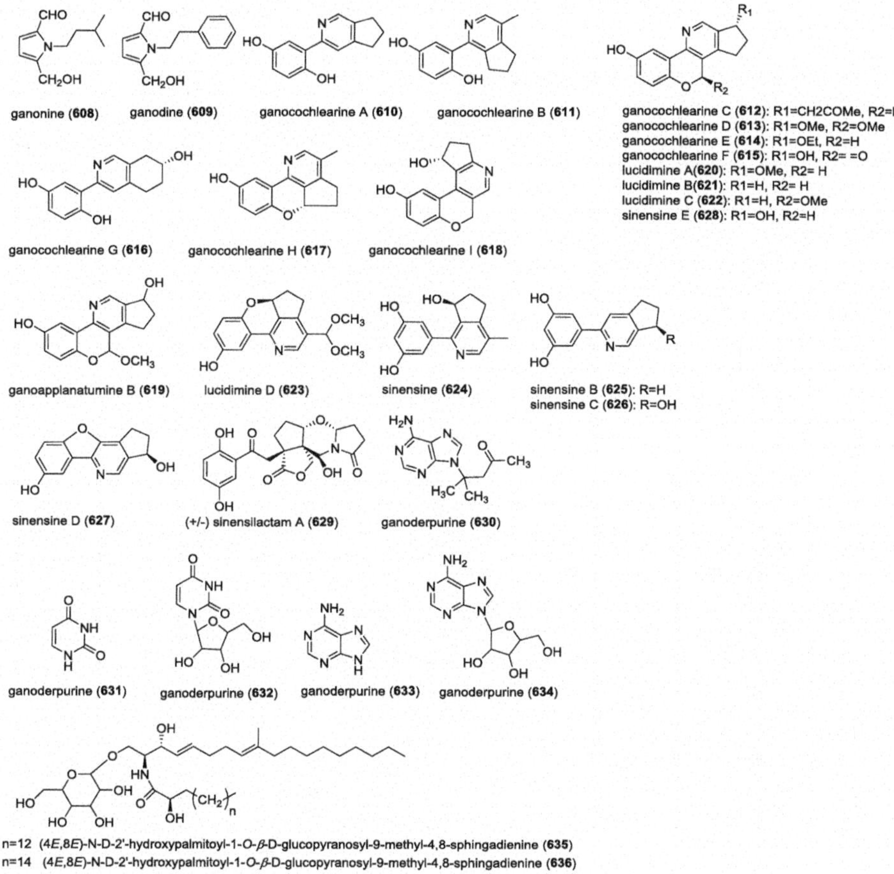

Fig. 3.10 Chemical structures of *Ganoderma* alkaloids

3.5.2.3 Ultrafiltration

Ultrafiltration is an effective method that combined the advantage of filtration and dialysis, which is used in desalination, concentration, and fractional separation of polysaccharide. The semipermeable membrane allows the small molecules to go through, but not for the big molecules. Thus, the compounds with different molecular weight can be separated. The type of ultrafiltration includes agitator, un-agitator, and hollow fiber ultrafiltration [176].

Table 3.11 The main structural characterization of polysaccharides isolated from *G. lucidum and G. tsugae*

No.	Origin	Part	Main glycosidic bonds	Sugar composition	Molecular weight	References
637	–	*G. lucidum* mycelium	–	**GLPI**: arabinose, rhamnose, xylose, mannose, glucose **GLPII**: arabinose, xylose, glucose **GLPIII**: arabinose, rhamnose, xylose, galactose, mannose, glucose **GLPIV**: arabinose, rhamnose, fucose, mannose, glucose	–	[160]
638	China	*G. lucidum* fruiting body	*β* (1–3)	Glucose	3.979 × 10^3 Da	[160]
639	China	*G. lucidum* fruiting body	*β* (1–3), (1–4), (1–6)	**GLP$_L$1**: glucose **GLP$_L$2**: glucose, galactose, mannose	**GLP$_L$1**: 5.2 × 10^3 Da **GLP$_L$2**: 15.4 × 10^3 Da	[160]
640	China	*G. lucidum* fruiting body	–	Mannose	–	[160]
641	China	*G. lucidum* fruiting body	–	Mannose	–	[160]
642	China	*G. lucidum* fruiting body	–	Rhamnose, xylose, fructose, galactose, mannose, glucose	5.85× 10^2 Da	[160]
643	Japan	*G. lucidum* fruiting body	*β* (1–3), (1–4), (1–6) *α* (1–4)	Glucose, xylose, arabinose	4× 10^4 Da	[160]
644	Japan	*G. lucidum* fruiting body	*β* (1–3), a few short (1–4)	Glucose, galactose, mannose, xylose, arabinose, fucose	–	[160]
645	Japan	*G. lucidum* mycelium	*β* (1–3)	Glucose	–	[160]
646	–	*G. lucidum* fruiting body	*β* (1–3), with *β* (1–6) branches	Glucose	**GLPO** < 1.2× 10^{4}Da; **GLPI** > 1.2× 10^{4}Da	[160]

(continued)

Table 3.11 (continued)

No.	Origin	Part	Main glycosidic bonds	Sugar composition	Molecular weight	References
647	China	*G. tsugae* mycelium	**GTM3(GTM4)**: *β* (1–3)	**GTM1 (GTM2)**: galactose, mannose **GTM3 (GTM4)**: glucose	**GTM1**: 6.28× 10^5Da; **GTM2**: 8.18× 10^5Da; **GTM3**: 4.65 × 10^6 Da; **GTM4**: 4.68× 10^6Da	[162]
648	Japan	*G. lucidum* fruiting body	*β* (1–3), (1–6)	Glucose	**Ganoderan B**: 7.41× 10^3Da; **Ganoderan C**: 5.81 × 10^3 Da	[163]
649	China	*G. lucidum* fruiting body	–	**GLB_6 (GLB_7)**: arabinose, xylose, galactose, mannose, glucose **GLB_{10} (GLC_1)**: rhamnose, arabinose, xylose, galactose, mannose, glucose	**GLB_6**: 8.8× 10^3Da; **GLB_7**: 9.0 × 10^3 Da; **GLB_{10}**: 6.8× 10^3Da; **GLC_1**: 5.7× 10^3Da	[164]
650	China	*G. lucidum* fruiting body	**TGLP-2 (TGLP-6)**: *β* (1–3), (1–6) **TGLP-3 (TGLP-7)**: *β* (1–3), *β* (1–4), (1–6)	**TGLP-2 (TGLP-3, TGLP-6)**: glucose **TGLP-7**: galactose	**TGLP-2**: 20.9× 10^4Da; **TGLP-3**: 4.5× 10^4Da; **TGLP-6**: 3.2× 10^4Da; **TGLP-7**: 10.0× 10^4Da	[165]
651	China	*G. lucidum* fruiting body	**PL-1**: *β* (1–4), (1–6) **PL-3**: *β* (1–3), (1–6) **PL-4**: *β* (1–3), (1–4), (1–6)	**PL-1**: glucose **PL-3**: glucose **PL-4**: glucose, mannose	**PL-1**: 8.3× 10^3Da; **PL-3**: 6.3× 10^4Da; **PL-4**: 2.0× 10^5Da	[166]
652	China	*G. lucidum* spores	*β* (1–3)	Glucose	**GSPL-I-1A**: 7.8× 10^5Da	[167]
653	China	*G. lucidum* fruiting body	–	Glucose, galactose, mannose	**GLIS**: -	[168]
654	China	*G. lucidum* fruiting body	–	Rhamnose, xylose, fructose, galactose, mannose, glucose	**GLPG**: 5.13× 10^5Da **GLPW**: 5.85× 10^5Da	[169]
655	China	*G. lucidum* fruiting body	*β* (1–3), (1–6)	Rhamnose, xylose, galactose, mannose, glucose	**GL-PP-3A**: 1.11× 10^4Da	[170]

(continued)

Table 3.11 (continued)

No.	Origin	Part	Main glycosidic bonds	Sugar composition	Molecular weight	References
656	China	*G. lucidum* mycelia	–	Rhamnose, xylose, galactose, glucose	**SeGLP-2B-1**: 1.06×10^6Da	[171]
657	China	*G. lucidum* fruiting body	–	–	**GL-IV-1**: 13.3×10^4Da; **S-GL**: 10.1×10^4Da; **CM-GL**: 6.3×10^4Da; **HE-GL**: 7.2×10^4Da; **HP-GL**: 5.1×10^4Da; **M-GL**: 14.1×10^4Da	[172]
658	China	*G. lucidum* fruiting body	β (1–3), (1–6)	Glucose	**GLP20**: 3.75×10^6Da	[173]

3.5.3 *The Determination of Relative Molecular Weights of Polysaccharides in* G. lucidum

To evaluate the correct molecular weights of polysaccharides, the polysaccharides should be superpurified. And some properties of polysaccharides are related to their relative molecular weights, such as the viscosity. With the increase of relative molecular weights, the viscosity of polysaccharides is also higher.

Electrospray ionization coupled with mass spectrometry (ESI-MS), high performance gel permeation chromatography (HPGPC), gel filtration, osmotic pressure, and vapor pressure osmometry are used in the evaluation of relative molecular weights [179, 180].

ESI-MS is an important analytical technique for MS analysis of polysaccharides, and it is extensively used in molecular weight determination of polysaccharides.

HPGPC is a type of size exclusion chromatography that separates analytes on the basis of size with high pressure. The high pressure allows the high speed of flow rate and fast separation of samples. One analysis can be finished in 30 mins.

3.5.4 The Structural Identification of Polysaccharides of G. lucidum

3.5.4.1 Methylation

All the hydroxyl groups in polysaccharides are methylated, and then the glycosidic bonds are hydrolyzed. After hydrolysis, the connection points between two monosaccharides are determined by the newly appeared hydroxy groups [181].

3.5.4.2 Smith Degradation

Periodic acid or periodate can split C–C bond of the monosaccharide ring. It is a quantitative reaction. When one C–C bond is breaking, one molar of periodic acid is required. According to the consumption of periodic acid, the number of monosaccharides and the type of glycosidic linkages are determined [182].

3.5.4.3 IR Spectrum

IR spectrum can determine the relative configurations of polysaccharides and some types of monosaccharides. The α-anomer and β-anomer show absorption peaks at 844 ± 8 cm^{-1} and 891 ± 7 cm^{-1}, respectively. Galactopyranoses show absorption peak at 875 cm^{-1}. Deoxyriboses, including fucose and rhamnose, display an absorption peak at 967 cm^{-1} [182].

3.5.4.4 NMR Spectrum

The relative configuration of monosaccharide in polysaccharides can be characterized by ^{1}H and ^{13}C NMR spectra. Most of the chemical shifts of monosaccharides in ^{1}H NMR spectrum are δ_H 3.5-4.4 ppm. However, the chemical shifts of anomeric hydrogen atoms are δ_H 4.4-5.8 ppm, which can be used to determine the relative configuration of monosaccharides. The chemical shifts of anomeric carbon atoms in ^{13}C NMR spectrum are δ_C 95-110 ppm, in which the chemical shifts of D-α-glucose and D-β-glucose are δ_C 97-101 and 103-106 ppm, respectively [183].

References

1. Gan KH, Fann YF, Hsu SH, Kou KW, Lin CN (1998) Mediation of the cytotoxicity of lanostanoids and steroids of *Ganoderma tsugae* through apoptosis and cell cycle. J Nat Prod 61(4):485–487

2. Niu XM, Li SH, Xiao WL, Sun HD, Che CT (2007) Two new lanostanoids from *Ganoderma resinaceum*. J Asian Nat Prod Res 9(7):659–664
3. Qiu J, Wang X, Song C (2016) Neuroprotective and antioxidant lanostanoid triterpenes from the fruiting bodies of *Ganoderma atrum*. Fitoterapia 109(4):75–79
4. Cheng CR, Yue QX, Wu ZY, Song XY, Tao SJ, Wu XH, Xu PP, Liu X, Guan SH, Guo DA (2010) Cytotoxic triterpenoids from *Ganoderma lucidum*. Phytochemistry 71(13):1579–1585
5. Jiao Y, Xie T, Zou LH, Wei Q, Qiu L, Chen LX (2016) Lanostane triterpenoids from *Ganoderma curtisii* and their NO production inhibitory activities of LPS-induced microglia. Bioorg Med Chem Lett 26(15):3556–3561
6. Hu LL, Ma QY, Huang SZ, Guo ZK, Ma HX, Guo JC, Dai HF, Zhao YX (2013) Three new lanostanoid triterpenes from the fruiting bodies of *Ganoderma tropicum*. J Asian Nat Prod Res 15(4):357–362
7. Shiao MS, Lin LJ, Yeh SF (1988) Triterpenes from *Ganoderma lucidum*. Phytochemistry 27(3):873–875
8. Hirotani M, Asaka I, Ino C, Furuya T, Shiro M (1987) Studies on the metabolites of higher fungi. Part 7. Ganoderic acid derivatives and ergosta-4,7,22-triene-3,6-dione from *Ganoderma lucidum*. Phytochemistry 26(10):2797–2803
9. Li YB, Liu RM, Zhong JJ (2013) A new ganoderic acid from *Ganoderma lucidum* mycelia and its stability. Fitoterapia 84:115–122
10. Nguyen VT, Tung NT, Cuong TD, Hung TM, Kim JA, Mi HW, Choi JS, Lee JH, Min BS (2015) Cytotoxic and anti-angiogenic effects of lanostane triterpenoids from *Ganoderma lucidum*. Phytochem Lett 12:69–74
11. Zhang SS, Wang YG, Ma QY, Huang SZ, Hu LL, Dai HF, Yu ZF, Zhao YX (2015) Three new lanostanoids from the mushroom *Ganoderma tropicum*. Molecules 20(2):3281–3289
12. Gonzalez AG, Leon F, Rivera A, Padron JI, Gonzalez-Plata J, Zuluaga JC (2002) New lanostanoids from the fungus *Ganoderma concinna*. J Nat Prod 65(3):417–421
13. Joseph S, Janardhanan KK, George V, Baby S (2011) A new epoxidic ganoderic acid and other phytoconstituents from *Ganoderma lucidum*. Phytochem Lett 4(3):386–388
14. Amen YM, Zhu Q, Afifi MS, Halim AF, Ashour A, Shimizu K (2016) New cytotoxic lanostanoid triterpenes from *Ganoderma lingzhi*. Phytochem Lett 17:64–70
15. Guan SH, Xia JM, Yang M, Wang XM, Liu X, Guo DA (2008) Cytotoxic lanostanoid triterpenes from *Ganoderma lucidum*. J Asian Nat Prod Res 10(8):695–700
16. Liu DZ, Zhu YQ, Li XF, Shan WG, Gao PF (2014) New triterpenoids from the fruiting bodies of *Ganoderma lucidum* and their bioactivities. Chem Biodivers 11(6):982–986
17. Tan Z, Zhao JL, Liu JM, Zhang M, Chen RD, Xie KB, Chen DW, Dai JG (2018) Lanostane triterpenoids and ergostane-type steroids from the cultured mycelia of *Ganoderma capense*. J Asian Nat Prod Res 20(9):844–851
18. Ma QY, Luo Y, Huang SZ, Guo ZK, Dai HF, Zhao YX (2013) Lanostane triterpenoids with cytotoxic activities from the fruiting bodies of *Ganoderma hainanense*. J Asian Nat Prod Res 15(11):1214–1219
19. Sato N, Zhang Q, Ma CM, Hattori M (2009) Anti-human immunodeficiency virus-1 protease activity of new lanostane-type triterpenoids from *Ganoderma sinense*. Chem Pharm Bull 57(10):1076–1080
20. Akihisa T, Tagata M, Ukiya M, Tokuda H, Suzuki T, Kimura Y (2005) Oxygenated lanostane-type triterpenoids from the fungus *Ganoderma lucidum*. J Nat Prod 68(4):559–563
21. Isaka M, Chinthanom P, Mayteeworakoon S, Laoteng K, Choowong W, Choeyklin R (2018) Lanostane triterpenoids from cultivated fruiting bodies of the basidiomycete *Ganoderma australe*. Nat Prod Res 32(9):1044–1049
22. Liu LY, Hui C, Chao L, Wang HQ, Kang J, Yan L, Chen RY (2014) Triterpenoids of *Ganoderma theaecolum* and their hepatoprotective activities. Fitoterapia 98:254–259
23. Lin LJ, Shiao MS (1988) Seven new triterpenes from *Ganoderma lucidum*. J Nat Prod 51(5):918–924

24. Isaka M, Chinthanom P, Sappan M, Danwisetkanjana K, Boonpratuang T, Choeyklin R (2016) Antitubercular lanostane triterpenes from cultures of the basidiomycete *Ganoderma* sp. BCC 16642. J Nat Prod 79(1):161–169
25. Zhao ZZ, Yin RH, Chen HP, Feng T, Li ZH, Dong ZJ, Cui BK, Liu JK (2015) Two new triterpenoids from fruiting bodies of fungus *Ganoderma lucidum*. J Asian Nat Prod Res 17(7):6
26. Ma QY, Huang SZ, Hu LL, Guo ZK, Dai HF, Zhao YX (2016) Two new tirucallane triterpenoids from the fruiting bodies of *Ganoderma tropicum*. Chem Nat Compound 52(4):656–659
27. Isaka M, Chinthanom P, Mayteeworakoon S, Laoteng K, Choowong W, Choeyklin R (2017) Lanostane triterpenoids from cultivated fruiting bodies of the basidiomycete *Ganoderma australe*. Phytochem Lett 21(9):251–255
28. Ha TB, Gerhauser C, Zhang WD, Ho-Chong-Line N, Fouraste I (2000) New lanostanoids from *Ganoderma lucidum* that induce NAD(P)H: quinone oxidoreductase in cultured hepalcic7 murine hepatoma cells. Planta Med 66(07):681–684
29. Mothana RAA, Ali NAA, Jansen R, Wegner U, Mentel R, Lindequist U (2003) Antiviral lanostanoid triterpenes from the fungus *Ganoderma pfeifferi*. Fitoterapia 74(1):177–180
30. Tung NT, Cuong TD, Hung TM, Lee JH, Mi HW, Choi JS, Kim J, Ryu SH, Min BS (2013) Inhibitory effect on NO production of triterpenes from the fruiting bodies of *Ganoderma lucidum*. Bioorg Med Chem Lett 23(5):1428–1432
31. Naoto S, Chao-Mei M, Katsuko K, Masao H (2009) Triterpene-farnesyl hydroquinone conjugates from *Ganoderma sinense*. J Nat Prod 72(5):958–961
32. Peng XR, Wang X, Zhou L, Hou B, Zuo ZL, Qiu MH (2015) Ganocochlearic acid A, a rearranged hexanorlanostane triterpenoid, and cytotoxic triterpenoids from the fruiting bodies of *Ganoderma cochlear*. RSC Adv 5(115):95212–95222
33. Kleinwachter P, Anh N, Kiet T, Schlegel B, Dahse H, Hartl A, Grafe U (2001) Colossolactones, new triterpenoid metabolites from a Vietnamese mushroom *Ganoderma colossum*. J Nat Prod 64(2):236
34. Kikuchi T, Kanomi S, Kadota S, Murai Y, Tsubono K, Ogita ZI (1986) Constitutents of the fungus *Ganoderma lucidum* (FR.) KARST. I.: structures of ganoderic acids C2, E, I, and K, lucidenic acid F and related compounds. Chem Pharma Bull 34(9):3695–3712
35. Lin KW, Chen YT, Yang SC, Wei BL, Hung CF, Lin CN (2013) Xanthine oxidase inhibitory lanostanoids from *Ganoderma tsugae*. Fitoterapia 89(1):231–238
36. Tsuyoshi NKO, Hiroji S, Sadao S (1988) Novel triterpenoids from the mycelial mat at the previous stage of fruiting of *Ganoderma lucidum*. Agric Bioi Chem 52(2):367–372
37. Nishitoba T, Sato H, Oda K, Sakamura S (1988) Novel triterpenoids and a steroid from the fungus *Ganoderma lucidum*. Agric Biol Chem 52(1):211–216
38. Komoda Y, Nakamura H, Ishihara S, Uchida M, kohda H, Yamasaki K (1985) Structures of new terpenoid constituents of *Ganoderma lucidum* (Fr.) Karst (Polyporaceae). Chem Pharma Bull, 33 (11): 4829–4835
39. Li P, Deng YP, Wei XX, Xu JH (2013) Triterpenoids from *Ganoderma lucidum* and their cytotoxic activities. Nat Prod Res 27(1):17–22
40. Niu X, Qiu M, Li Z, Lu Y, Cao P, Zheng Q (2004) Two novel 3,4-seco-trinorlanostane triterpenoids isolated from *Ganoderma fornicatum*. Tetrahedron Lett 45(14):2989–2993
41. Peng XR, Liu JQ, Xia JJ, Yang YH, Qiu MH (2012) Two new triterpenoids from *Ganoderma cochlear*. Chin Trad Herb Drug 43(6):1045–1049
42. Ma K, Ren J, Han J, Bao L, Li L, Yao Y, Sun C, Zhou B, Liu H (2014) Ganoboninketals A-C, Antiplasmodial 3,4-seco-27-norlanostane triterpenes from *Ganoderma boninense* Pat. J Nat Prod 77(8):1847–1852
43. Ma K, Li L, Bao L, He L, Sun C, Zhou B, Si S, Liu H (2015) Six new 3,4-seco-27-norlanostane triterpenes from the medicinal mushroom *Ganoderma boninense* and their antiplasmodial activity and agonistic activity to LXRβ. Tetrahedron 71(12):1808–1814
44. Morigiwa A, Kitabatake K, Fujimoto Y, Ikekawa N (1986) Angiotensin converting enzyme-inhibitory triterpenes from *Ganoderma lucidim*. Chem Pharma Bull 34(7):3025–3028

45. Peng XR, Liu JQ, Wang CF, Li XY, Shu Y, Zhou L, Qiu MH (2014) Hepatoprotective effects of triterpenoids from *Ganoderma cochlear*. J Nat Prod 77(4):737–743
46. Li W, Lou LL, Zhu JY, Zhang JS, Liang AA, Bao JM, Tang GH, Yin S (2016) New lanostane-type triterpenoids from the fruiting body of *Ganoderma hainanense*. Fitoterapia 115:24–30
47. Kohda H, Tokumoto W, Sakamoto K, Fujii M, Hirai Y, Yamasaki K, Komoda Y, Nakamura H, Ishihara S, Uchida M (1985) The biologically active constituents of *Ganoderma lucidum* (Fr.) Karst. Histamine release-inhibitory triterpenes. Chem Pharma Bull 33(4):1367
48. Fatmawati S, Shimizu K, Kondo R (2010) Ganoderic acid Df, a new triterpenoid with aldose reductase inhibitory activity from the fruiting body of *Ganoderma lucidum*. Fitoterapia 81(8):1033–1036
49. Wang F, Cai H, Yang J, Zhang Y, Hou C, Liu J, Zhao M (1997) Studies on ganoderic acid, a new constituent from the fruiting body of *Ganoderma lucidum* (FR.) karst. Acta Pharm Sin 32(6):447–450
50. Luo J, Zhao YZ (2002) A new lanostane-type triterpene from the fruiting bodies of *Ganoderma lucidum*. J Asian Nat Prod Res 4(2):129–134
51. Nishitoba T, Sato H, Shirasu S, Sakamura S (1987) Novel triterpenoids from the mycelial mat at the previous stage of fruiting of *Ganoderma lucidum*. Agric Biol Chem 51(2):619–622
52. Toth JO, Luu B, Beck JP, Ourisson G (1983) Chemistry and biochemistry of oriental drugs. Part IX. Cytotoxic triterpenes from *Ganoderma lucidum* (Polyporaceae): structures of ganoderic acids U-Z. J Chem Res Synop 12:299
53. Nishitoba T, Sato H, Sadao S (1987) Novel mycelial components ganoderic acid Mg Mh Mi Mj and Mk from the fungus *Ganoderma lucidum*. Agric Biol Chem 51(4):1149–1153
54. Hirotani M, Ino C, Furuya T, Shiro M (1986) Ganoderic acids T, S and R, new triterpenoids from the cultured mycelia of *Ganoderma lucidum*. Chem Pharma Bull 34(5):2282–2285
55. Hirotani M, Asaka I, Furuya T (1990) Studies on the metabolites of higher fungi. 9. Investigation of the biosynthesis of 3α-hydroxy triterpenoids,. ganoderic acids T and S, by application of a feeding experiment using [1,2-13C2]acetate. J Chem Soc Perkin Trans 1(10):2751–2754
56. Li CH, Chen PY, Chang UM, Kan LS, Fang WH, Tsai KS, Lin SB (2005) Ganoderic acid X, a lanostanoid triterpene, inhibits topoisomerases and induces apoptosis of cancer cells. Life Sci 77(3):252–265
57. Liu LY, Yan Z, Kang J, Chen RY, Yu DQ (2017) Three new triterpenoids from *Ganoderma theaecolum*. J Asian Nat Prod Res 19(9):847–853
58. Jo T, Luu B, Ourisson G (1983) Les acides ganoderiques tàz: triterpenes cytotoxiques de *Ganoderma lucidum* (Polyporacée). Tetrahedron Lett 24(10):1081–1084
59. El-Mekkawy S, Meselhy MR, Nakamura N, Tezuka Y, Hattori M, Kakiuchi N, Shimotohno K, Kawahata T, Otake T (1998) Anti-HIV-1 and anti-HIV-1-protease substances from *Ganoderma lucidum*. Phytochemistry 49(6):1651
60. Min BS, Nakamura N, Miyashiro H, Bae KW, Hattori M (1998) Triterpenes from the Spores of *Ganoderma lucidum* and their inhibitory activity against HIV-1 Protease. Chem Pharm Bull 46(10):1607–1612
61. Min BS, Gao JJ, Nakamura N, Hattori M (2000) Triterpenes from the spores of *Ganoderma lucidum* and their cytotoxicity against Meth-A and LLC tumor cells. Chem Pharm Bull 48(7):1026–1033
62. Hiroji STN, Sachiko S, Kyoko ODA, Sadao S (1986) Ganoderiol A and B, new triterpenoids from the fungus *Ganoderma lucidum* (Reishi). Agric Biol Chem 50(11):2887–2890
63. Zhao XR, Huo XK, Dong PP, Wang C, Huang SS, Zhang BJ, Zhang HL, Deng S, Liu KX, Ma XC (2015) Inhibitory effects of highly oxygenated lanostane derivatives from the fungus *Ganoderma lucidum* on P-Glycoprotein and α-Glucosidase. J Nat Prod 78(8):1868–1876
64. Waranya L, Kwanjai K, Somdej K, Palangpon K, Samran P, Phoutthasone S, Kasem S (2014) Triterpene lactones from cultures of *Ganoderma* sp. KM01. J Nat Prod 77(7):1545–1553
65. Wang FS, Chai H, Yang JS, Zhang JM, Zhao YJ (1997) Triterpenoids from the fruiting body of *Ganoderma lucidum*. J Chin Pharm Sci 6(4):20–25

66. Lian C, Wang CF, Xiao Q, Xiao L, Xu Y, Liu J (2017) The triterpenes and steroids from the fruiting body *Ganoderma duripora*. Biochem System Eco 73:50–53
67. Qiao Y, Zhang XM, Qiu MH (2007) Two novel lanostane triterpenoids from *Ganoderma Sinense*. Molecules 12(8):2038
68. Wang K, Bao L, Xiong W, Ma K, Han J, Wang W, Yin W, Liu H (2015) Lanostane Triterpenes from the tibetan medicinal mushroom *Ganoderma leucocontextum* and their inhibitory effects on HMG-CoA reductase and α-Glucosidase. J Nat Prod 78(8):1977–1989
69. Tsuyoshi N, Hiroji S, Sadao S (1986) New terpenoids ganolucidic acid D ganoderic acid L lucidone C and lucidenic acid G from the Fungus *Ganoderma lucidum*. Agric Biol Chem 50(3):809–811
70. Chen B, Jin T, Zhang J, Kai W, Li L, Bo Y, Li B, Liu H (2017) Triterpenes and meroterpenes from *Ganoderma lucidum* with inhibitory activity against HMGs reductase, aldose reductase and α-glucosidase. Fitoterapia 120:6–16
71. Isaka M, Chinthanom P, Kongthong S, Srichomthong K, Choeyklin R (2013) Lanostane triterpenes from cultures of the Basidiomycete *Ganoderma orbiforme* BCC 22324. Phytochemistry 87(3):133–139
72. Wang CF, Liu JQ, Yan YX, Chen JC, Lu Y, Guo YH, Qiu MH (2010) Three new triterpenoids containing four-membered ring from the Fruiting Body of *Ganoderma sinense*. Org Lett 12(8):1656–1659
73. Chen RY, Yu DQ (1991) Application of 2d NMR techniques in the structure determination of ganosporelactone A and B. Acta Pharm Sin 26(6):430–436
74. Ma L, Wu F, Chen R (2003) Analysis of triterpene constituents from *Ganoderma lucidum*. Acta Pharm Sin 38(1):50–52
75. Pu DB, Zheng X, Gao JB, Zhang XJ, Qi Y, Li XS, Wang YM, Li XN, Li XL, Wan CP (2017) Highly oxygenated lanostane-type triterpenoids and their bioactivity from the fruiting body of *Ganoderma gibbosum*. Fitoterapia 119:1–7
76. Zhao ZZ, Chen HP, Li ZH, Dong ZJ, Bai X, Zhou ZY, Feng T, Liu JK (2016) Leucocontextins A–R, lanostane-type triterpenoids from *Ganoderma leucocontextum*. Fitoterapia 109:91–98
77. Zhao ZZ, Chen HP, Huang Y, Li ZH, Zhang L, Feng T, Liu JK (2016) Lanostane triterpenoids from fruiting bodies of *Ganoderma leucocontextum*. Nat Prod Biopro 6(2):103–109
78. Gao JJ, Min BS, Ahn EM, Nakamura N, Lee HK, Hattori M (2002) New triterpene aldehydes, lucialdehydes A-C, from *Ganoderma lucidum* and their cytotoxicity against murine and human tumor cells. Chem Pharma Bull 50(6):837–840
79. Tsuyoshi N, Takanori K, Hirokazu K (1985) New bitter C27 and C30 terpenoids from the fungus *Ganoderma lucidum* Reishi. Agric Biol Chem 49(6):1793–1798
80. Kikuchi T, Kanomi S, Murai Y, Kadota S, Tsubono K, Ogita ZI (1986) Constituents of the fungus *Ganoderma lucidum* (FR.) KARST. III.: structures of ganolucidic acids A and B, new lanostane-type triterpenoids. Chem Pharma Bull 34(10):4030–4036
81. Wu T, Shi L, Kuo S (2001) Cytotoxicity of *Ganoderma lucidum* triterpenes. J Nat Prod 64(8):1121–1122
82. Mizushina Y, Takahashi N, Hanashima L, Koshino H, Esumi Y, Uzawa J, Sugawara F, Sakaguchi K (1999) Lucidenic acid O and lactone, new terpene inhibitors of eukaryotic DNA polymerases from a basidiomycete, *ganoderma lucidum*. Bioorg Med Chem 7(9):2047–2052
83. Iwatsuki K, Akihisa T, Tokuda H, Ukiya M, Oshikubo M, Kimura Y (2003) Lucidenic acids P and Q, methyl lucidenate P, and other triterpenoids from the fungus *Ganoderma lucidum* and their inhibitory effects on Epstein-Barr virus activation. J Nat Prod 66(12):1582–1585
84. Liu C, Li BM, Kang J, Wang HQ, Chen RY (2013) One new triterpenoid from fruiting body of *Ganoderma lucidum*. Acta Pharm Sin 48(9):1450–1452
85. Amen YM, Zhu Q, Tran HB, Afifi MS, Halim AF, Ashour A, Mira A, Shimizu K (2016) Lucidumol C, a new cytotoxic lanostanoid triterpene from *Ganoderma lingzhi* against human cancer cells. J Nat Med 70(3):1–6

86. Kikuchi T, Matsuda S, Kadota S, Murai Y, Ogita Z (1985) Ganoderic acid D, E, F, and H and lucidenic acid D, E, and F, new triterpenoids from *Ganoderma lucidum*. Chem Pharma Bull 33(6):2624–2627
87. Tsuyoshi N (1987) Triterpenenoids from the fungus *Ganoderma lucidum*. Phytochemistry 26(6):1777–1784
88. Tsuyoshi N (1985) New Terpenoids from *Ganoderma lucidum* and their bitterness. Agric Biol Chem 49(5):1547–1549
89. Guo PX, Wang XL, Cheng YX, Pei MR (2016) Three new terpenoids from *Ganoderma lucidum*. Nat Prod Res Dev 28(1):1–4
90. Liu C, Chen RY (2010) A new triterpene from fungal fruiting bodies of *Ganoderma sinense*. Chin Trad Herb Drugs 41(1):8–11
91. Lin CN, Fann YF, Chung MI (1997) Steroids of formosan *Ganoderma tsugae*. Phytochemistry 46(6):1143–1146
92. Su H, Fann YF, Chung MI, Won SJ, Lin C (2000) New lanostanoids of *Ganoderma tsugae*. J Nat Prod 63(4):514–516
93. Xu HY, Hou MN, Liu J (2007) Study on extraction technology of total triterpenoids from *Ganoderma lucidum*. J ShaanXi College Trad Chin Med 30(3):43–44
94. Gong XF, Xie MY, Chen Y (2006) Spectrophotometric determination of triterpenoids and triterpenoid saponins in *Gendarme atrum*. Nat Prod Res Develop 18(5):825–829
95. Chen Y, Xie MY, Gong XF (2007) Microwave-assisted extraction used for the isolation of total triterpenoid saponins from *Ganoderma atrum*. J Food Eng 81(1):162–170
96. Zhang J, Duan JC, Liang Z, Zhang WB, Zhang LH, Huo YS, Zhang YK (2006) Analysis of triterpenoids in fruiting bodies of *Ganoderma lucidum* with off-line supercritical fluid extraction-high performance liquid chromatography system. Chin J Analy Chem 34(4):477–450
97. Mothana RAA, Jansen R, Julich WD, Lindequist U (2000) Ganomycins A and B, new antimicrobial farnesyl hydroquinones from the basidiomycete *Ganoderma pfeifferi*. J Nat Prod 63:416–418
98. Peng XR, Li L, Wang X, Zhu GL, Zhou L, Qiu MH (2016) Antioxidant farnesylated hydroquinones from *Ganoderma capense*. Fitoterapia 111:18–23
99. Wang K, Bao L, Ma K, Zhang JJ, Chen BS, Han JJ, Ren JW, Luo HJ, Liu HW (2017) A novel class of α-glucosidase and HMG-CoA reductase inhibitors from *Ganoderma leucocontextum* and the anti-diabetic properties of ganomycin I in KK-Ay mice. Eur J Med Chem 127:1035–1046
100. Chen BS, Tian J, Zhang JJ, Wang K, Liu L, Yang B, Bao L, Liu HW (2017) Triterpenes and meroterpenes from *Ganoderma lucidum* with inhibitory activity against HMGs reductase, aldose reductase and α-glucosidase. Fitoterapia 120:6–16
101. Peng XR, Liu JQ, Wang CF, Han ZH, Shu Y, Li XY, Zhou L, Qiu MH (2015) Unusual prenylated phenols with antioxidant activities from *Ganoderma cochlear*. Food Chem 171:251–257
102. Wang XL, Wu ZH, Di L, Zhou FJ, Yang YM, Cheng YX (2019) Renoprotective meroterpenoids from the fungus *Ganoderma cochlear*. Fioterapia 132:88–93
103. Peng XR, Wang X, Chen L, Yang H, Lei L, Lu SY, Zhou L, Qiu MH (2018) Racemic meroterpenoids from *Ganoderma cochlear*. Fitoterapia 127:286–292
104. Luo Q, Wang XL, Di L, Yan YM, Lu Q, Yang XH, Hu DB, Cheng YX (2015) Isolation and identification of renoprotective substances from the mushroom *Ganoderma lucidum*. Tetrahedron 71(5):840–845
105. Luo Q, Yang XH, Yang ZL, Tu ZC, Cheng YX (2016) Miscellaneous meroterpenoids from *Ganoderma applanatum*. Tetrahedron 72(30):4564–4574
106. Luo Q, Li MK, Luo JF, Tu ZC, Cheng YX (2018) COX-2 and JAK3 inhibitory meroterpenoids from the mushroom *Ganoderma theaecolum*. Tetrahedron 74:4259–4265
107. Zhang JJ, Dong Y, Qin FY, Cheng YX (2019) Australeols A–F, neuroprotective meroterpenoids from *Ganoderma australe*. Fitoterapia 134:250–255
108. Dou M, Li RT, Chen YX (2016) Minor compounds from fungus *Ganoderma cochlear*. Chin Herb Med 8(1):85–88

109. Chen XQ, Chen LX, Li SP, Zhao J (2017) Meroterpenoids from the fruiting bodies of higher fungus *Ganoderma resinaceum*. Phytochem Lett 22:214–218
110. Huang SZ, Cheng BH, Ma QY, Wang Q, Kong FD, Dai HF, Qiu SQ, Zheng PY, Liu ZQ, Zhao YX (2016) Anti-allergic prenylated hydroquinones and alkaloids from the fruiting body of *Ganoderma calidophilum*. RSC Adv 6:21139
111. Niu XM, Li SH, Sun HD, Che CT (2006) Prenylated phenolics from *Ganoderma fornicatum*. J Nat Prod 69:1364–1365
112. Cao WW, Luo Q, Chen YX, Wang SM (2016) Meroterpenoid enantiomers from *Ganoderma sinensis*. Fitoterapia 110:110–115
113. Yan YM, Wang XL, Luo Q, Jiang LP, Yang CP, Hou B, Zuo ZL, Chen YB, Cheng YX (2015) Metabolites from the mushroom *Ganoderma lingzhi* as stimulators of neural stem cell proliferation. Phytochemistry 114:155–162
114. Wang XF, Yan YM, Wang XL, Ma XJ, Fu XY, Cheng YX (2015) Two new compounds from *Ganoderma lucidum*. J Asian Nat Prod Res 17(4):329–332
115. Luo Q, Tu ZC, Cheng YX (2017) Two rare meroterpenoidal rotamers from *Ganoderma applanatum*. RSC Adv 7(6):3413–3418
116. Li CG, Luo Q, Guo PX, Chen LL, Cheng YX (2016) Petchiethers A and B, novel meroterpenoids with a 14- or 15-membered ring from *Ganoderma petchii*. Phytochem Lett 18:14–18
117. Luo Q, Tu ZC, Yang ZL, Cheng YX (2018) Meroterpenoids from the fruiting bodies of *Ganoderma theaecolum*. Fitoterapia 125:273–280
118. Zhou FJ, Wang XL, Wang SM, Cheng YX (2015) A new meroterpenoid from *Ganoderma lucidum*. Nat Prod Res Dev 27(1):22–25
119. Gao QL, Guo PX, Luo Q, Yan H, Cheng YX (2015) Petchienes A-E, meroterpenoids from *Ganoderma petchii*. Nat Prod Commun 10(12):2019–2022
120. Luo Q, Wei XY, Yang J, Luo JF, Liang R, Tu ZC, Cheng YX (2017) Spiro meroterpenoids from *Ganoderma applanatum*. J Nat Prod 80(1):61–70
121. Zhang JJ, Ma K, Chen HY, Wang K, Xiong WP, Bao L, Liu HW (2017) A novel polycyclic meroterpenoid with aldose reductase inhibitory activity from medicinal mushroom *Ganoderma leucocontextum*. J Antibiot (Tokyo) 70(8):915–917
122. Yan YM, Wang XL, Zhou LL, Zhou FJ, Li R, Tian Y, Zuo ZL, Fang P, Chung ACK, Hou FF, Cheng YX (2015) Lingzhilactones from *Ganoderma lingzhi* ameliorate adriamycin-induced nephropathy in mice. J Ethnopharmacol 176:385–393
123. Peng XR, Lu SY, Shao LD, Zhou L, Qiu MH (2018) Structural elucidation and biomimetic synthesis of (±)-cochlactone A with anti-inflammatory activity. J Org Chem 83:5516–5522
124. Yan YM, Ai J, Zhou LL, Chung ACK, Li R, Nie J, Fang P, Wang XL, Luo J, Hu Q, Hou FF, Cheng YX (2013) Lingzhiols, unprecedented rotary door-shaped meroterpenoids as potent and selective inhibitors of p-Smad3 from *Ganoderma lucidum*. Org Lett 15(21):5488–5491
125. Peng XR, Liu JQ, Wan LS, Li XN, Yan YY, Qiu MH (2014) Four new polycyclic meroterpenoids from *Ganoderma cochlear*. Org Lett 16:5262–5265
126. Dou M, Di L, Zhou LL, Yan YM, Wang XL, Zhou FJ, Yang ZL, Li RT, Hou FF, Cheng YX (2014) Cochlearols A and B, polycyclic meroterpenoids from the fungus *Ganoderma cochlear* that have renoprotective activities. Org Lett 16(23):6064–6067
127. Luo Q, Di L, Yang XH, Cheng YX (2016) Applanatumols A and B, meroterpenoids with unprecedented skeletons from *Ganoderma applanatum*. RSC Adv 6(51):45963–45967
128. Li L, Li H, Peng XR, Hou B, Yu MY, Dong JR, Li XN, Zhou L, Yang J, Qiu MH (2016) (±)-Ganoapplanin, a pair of polycyclic meroterpenoid enantiomers from *Ganoderma applanatum*. Org Lett 18(23):6078–6081
129. Luo Q, Di L, Dai WF, Lu Q, Yan YM, Yang ZL, Li RT, Cheng YX (2015) Applanatumin A, a new dimeric meroterpenoid from *Ganoderma applanatum* that displays potent antifibrotic activity. Org Lett 17(5):1110–1113
130. Zhou FJ, Nian Y, Yan YM, Gong Y, Luo Q, Zhang Y, Hou B, Zuo ZL, Wang SM, Jiang HH, Yang J, Cheng YX (2015) Two new classes of t-type calcium channel inhibitors with new chemical scaffolds from *Ganoderma cochlear*. Org Lett 17(12):3082–3085

131. Chen HP, Zhao ZZ, Zhang Y, Bai X, Zhang L, Liu JK (2016) (+)- and (−)-ganodilactone, a pair of meroterpenoid dimers with pancreatic lipase inhibitory activities from the macromycete *Ganoderma leucocontextum*. RSC Adv 6(69):64469–64473
132. Qin FY, Yang YM, Tu ZC, Cheng YX (2018) Meroterpenoid dimers from *Ganoderma cochlear* and their cytotoxic and COX-2 inhibitory activities. Fitoterapia 129:167–172
133. Wang XL, Zhou FJ, Dou M, Yan YM, Wang SM, Di L, Cheng YX (2016) Cochlearoids F–K: phenolic meroterpenoids from the fungus *Ganoderma cochlear* and their renoprotective activity. Bioorg Med Chem Lett 26(22):5507–5512
134. Qin FY, Yang YM, Tu ZC, Cheng YX (2018) (±) Gancochlearols A and B: cytotoxic and COX-2 inhibitory meroterpenoids from *Ganoderma cochlear*. Nat Prod Res:1. https://doi.org/10.1080/14786419.2018.1531859
135. Peng XR, Qiu MH (2018) Meroterpenoids from Ganoderma species: a review of last five years. Nat Prod Bioprospect 8:137–149
136. Bao FY, Yang KY, Wu CR, Gao SY, Wang PH, Chen LX, Li H (2018) New natural inhibitors of hexokinase 2 (HK2): steroids from *Ganoderma Sinense*. Fitoterapia 125:123–129
137. Chen RY, Wang YH, Yu DQ (1991) Studies on the chemical constituents of the spores from *Ganoderma lucidum*. Acta Bot Sin 33(1):65–68
138. Chen XQ, Chen LX, Li SP, Zhao J (2017) A new nortriterpenoid and an ergostane-type steroid from the fruiting bodies of the fungus *Ganoderma resinaceum*. J Asian Nat Prod Res 19(12):1–6
139. Gan KH, Kuo SH, Lin CN (1998) Steroidal constituents of *Ganoderma applanatum* and *Ganoderma neo-japonicum*. J Nat Prod 61(11):1421–1422
140. Thang TD, Kuo PC, Hwang TL, Yang ML, Ngoc NTB, Han TTN, Lin CW, Wu TS (2013) Triterpenoids and steroids from *Ganoderma mastoporum* and their inhibitory effects on superoxide anion generation and elastase release. Molecules 18(11):14285–14292
141. Ko HH, Hung CF, Wang JP, Lin CN (2008) Antiinflammatory triterpenoids and steroids from *Ganoderma lucidum* and *G. tsugae*. Phytochemistry 69(1):234–239
142. Su HJ, Fann YF, Chung MI, Won SJ, Lin CN (2000) New lanostanoids of *Ganoderma tsugae*. J Nat Prod 63(4):514–516
143. Lin CN, Tome WP, Won SJ (1991) Novel cytotoxic principles of Formosan *Ganoderma lucidum*. J Nat Prod 54(4):998–1002
144. Seo HW, Hung TM, Na MK, Jung HJ, Kim JC, Choi JS, Kim JH, Lee HK, Lee IS, Bae KH, Hatorri M, Min BS (2009) Steroids and triterpenes from the fruit bodies of *Ganoderma lucidum* and their anti-complement activity. Arch Pharm Res 32(11):1573–1579
145. Liu C, Chen RY (2010) A new sterol from the fruiting bodies of *Ganoderma sinense* Zhao, Xue et Zhang. Chin Pharm J 45(6):413–415
146. Wu SH, Huang R, Miao CP, Chen YW (2013) Two new steroids from an endophytic *Fungus phomopsis* sp. Chem Biodivers 10(7):276–1283
147. Hirotani M, Asaka I, Ino C, Furuya T, Shiro M (1987) Ganoderic acid derivatives and ergosta-4,7,22-triene-3,6-dione from *Ganoderma lucidum*. Phytochemistry 26(10):2797–2803
148. Tan Z, Zhao JL, Liu JM, Zhang M, Chen RD, Xie KB, Chen DW, Dai JG (2015) Lanostane triterpenoids and ergostane-type steroids from the cultured mycelia of *Ganoderma capense*. J Asian Nat Prod Res 17:1–8
149. Liu C, Wang HQ, Li BM, Chen RY (2007) Studies on chemical constituents from the fruiting bodies of *Ganoderma sinense* Zhao, Xue et Zhang. Chin J Chin Materia Medica 32(3):235–237
150. Yu JG, Chen RY, Yao ZX, Zhai YF, Yang SL, Ma JL (1990) Studies on constituents of *Ganoderma capense* IV. The chemical structures of ganonine, ganodine and ganoderpurine. Acta Pharm Sin 25(8):612–616
151. Tian L, Wang XL, Wang YZ, Cheng YX (2015) Two new alkaloids from *Ganoderma cochlear*. Nat Prod Res Dev 27:1325–1328
152. Wang XL, Dou M, Luo Q, Cheng LZ, Yan YM, Li RT, Cheng YX (2017) Racemic alkaloids from the fungus *Ganoderma cochlear*. Fitoterapia 116:93–98

153. Luo Q, Yang XH, Yang ZL, Tu ZC, Cheng YX (2016) Miscellaneous meroterpenoids from *Ganoderma applanatum*. Tetrahedron 72:1–11
154. Zhao ZZ, Chen HP, Feng T, Li ZH, Dong ZJ, Liu JK (2015) Lucidimine A-D, four new alkaloids from the fruiting bodies of *Ganoderma lucidum*. J Asian Nat Prod Res 17(12):1160–1165
155. Liu JQ, Wang CF, Peng XR, Qiu MH (2011) New alkaloids from the fruiting bodies of *Ganoderma sinense*. Nat Prod Bioprospect 1(2):93–96
156. Liu C, Zhao F, Chen RY (2010) A novel alkaloid from the fruiting bodies of *Ganoderma sinense* Zhao, Xu et Zhang. Chin Chem Lett 21(2):197–199
157. Luo Q, Tian L, Di L, Yan YM, Wei XY, Wang XF, Cheng YX (2015) (±)-Sinensilactam A, a pair of rare hybrid metabolites with Smad3 phosphorylation inhibition from *Ganoderma sinensis*. Org Lett 17(6):1565–1568
158. Shimizu A, Yano T, Saito Y, Inada Y (1985) Isolation of an inhibitor of platelet aggregation from a fungus, *Ganoderma lucidum*. Chem Pharm Bull 33(7):3012–3015
159. Yoshiyuki M, Linda H, Toyofumi Y, Masaharu T, Fumio S, Mineo S, Akio M, Shonen Y, Kengo S (1998) A mushroom fruiting body-inducing substance inhibits activities of replicative DNA polymerases. Biochem Bioph Res Co 249(1):17–22
160. Ferreira IC, Heleno SA, Reis FS, Stojkovic D, Queiroz MJ, Vasconcelos MH, Sokovic M (2015) Chemical features of *Ganoderma* polysaccharides with antioxidant, antitumor and antimicrobial activities. Phytochemistry 114:38–55
161. Liu Y, Tang Q, Zhang J, Xia Y, Yang Y, Wu D, Fan H, Cui SW (2018) Triple helix conformation of *β*-D-glucan from *Ganoderma lucidum* and effect of molecular weight on its immunostimulatory activity. Int J Biol Macromol 114:1064–1070
162. Peng Y, Zhang L, Zeng L, Kennedy J (2005) Structure and antitumor activities of the water-soluble polysaccharides from *Ganoderma tsugae* mycelium. Carbohyd Polym 59:385–392
163. Tomoda M, Gonda R, Kasahara Y, Hikino H (1986) Glycan structures of ganoderans B and C, hypoglycemic glycans of *Ganoderma lucidum* fruit bodies. Phytochemistry 25(12):2817–2820
164. Chen Q, Li R, He Y, Lei L, Lin Z, Shen C, Lu C (1993) Studies on anti-aging polysaccharides GLB GLC of *Ganoderma lucidum*. Beijing Yike Daxue Xuebao 25(4):303–305
165. Li T, He Y, Li R (1997) The study of polysaccharides of *Ganoderma lucidum* from Tai Mountain. China J Chinese Matera Medica 22(8):487–489
166. Bao XF, Wang XS, Dong Q, Fang JN, Li XY (2002) Structural features of immunologically active polysaccharides from *Ganoderma lucidum*. Phytochemistry 59(2):175–181
167. Bao XF, Zhen Y, Ruan L, Fang JN (2002) Purification, characterization, and modification of T lymphocyte-stimulating polysaccharide from spores of *Ganoderma lucidum*. Chem Pharm Bull 50(5):623–629
168. Zhang J, Tang Q, Zimmerman-Kordmann M, Reutter W, Fan H (2010) Activation of B lymphocytes by GLIS, a bioactive proteoglycan from *Ganoderma lucidum*. Life Sci 87(19–22):628–637
169. Lin S, Wang S, Lin Z, Lin Y (2003) Isolation and identification of active compounds of *Ganoderma lucidum* cultivated with grass and wood log I. Extraction, purification and characterization of glycopeptide. Chin Trad Herb Drug 10:872–874
170. Wang S, Kan D, Lin S, Lin Z (2007) Isolation, purification and structural analysis of GL-PP-3A, an active polysaccharide peptide from *Ganoderma lucidum*. Acta Pharm Sin 42(10):1058–1061
171. Shang D, Zhang J, Wen L, Li Y, Cui Q (2009) Preparation, Characterization, and antiproliferative activities of the Se-containing polysaccharide SeGLP-2B-1 from Se-enriched *Ganoderma lucidum*. J Agric Food Chem 57:7737–7742
172. Wang J, Zhang L (2009) Structure and chain conformation of five water-soluble derivatives of a β-D-glucan isolated from *Ganoderma lucidum*. Carbohydr Res 344:105–112
173. Liu Y, Zhang J, Tang Q, Yang Y, Guo Q, Wang Q, Wu D, Cui SW (2014) Physicochemical characterization of a high molecular weight bioactive *β*-D-glucan from the fruiting bodies of *Ganoderma lucidum*. Carbohydr Polym 101:968–974

174. Shi L (2016) Bioactivities, isolation and purification methods of polysaccharides from natural products: a review. Int J Biol Macromol 92:37–48
175. Wang M, Fang W, Li X (2014) Research progress in isolation, purification, physio-chemical properties and bioactivities of tea seed polysaccharide. Food Sci:280–285
176. Xue D, Huang D, Huang G, Sun L (2014) The advances of the isolation and separation of polysaccharides in plants. J Chin Med Mat 37(1):157–161
177. Momoi T, Ando S, Magai Y (1976) High resolution preparative column chromatographic system for gangliosides using DEAE-Sephadex and a new porus silica, latrobeads. Biochim Biophys Acta 441(3):488–497
178. Shang J, Cui Q, Tian X (2002) The purification and the analysis of molecular composition. Edib Fungi China 21(3):35–37
179. Gangula S, Nimtz M, Mischnick P (2016) Relative ion intensities of maltooligosaccharide ethers in electrospray ionization ion trap mass spectrometry: a quantitative evaluation. Int J Mass Spectrom 402:57–65
180. Chen R, Meng F, Liu Z, Chen R, Zhang M (2010) Characterization and anti-allergic effect of a polysaccharide from the flower buds of *Lonicera japonica*. Carbohydr Polym 80:845–851
181. Tian J, Che H, Ha D, Wei Y, Zheng S (2012) Characterization and anti-allergic effect of a polysaccharide from the flower buds of *Lonicera japonica*. Carbohydr Polym 90:1642–1647
182. Zhang Y, Wang Z, Li D, Zang W, Zhu H, Wu P, Mei Y, Liang Y (2016) A polysaccharide from *Antrodia cinnamomea* mycelia exerts antitumor activity through blocking of TOP1/TDP1-mediated DNA repair pathway. Int J Biol Macromol 92:37–48
183. Ye L, Zhang J, Pan Y (2017) The NMR technology applied in the structural elucidation of polysaccharides in edible and medicinal fungi. Acta Edulis Fungi 14(4):68–75

Chapter 4
Polysaccharide of *Ganoderma* and Its Bioactivities

Jingsong Zhang, Yanfang Liu, Qingjiu Tang, Shuai Zhou, Jie Feng, and Hongyu Chen

Abstract *Ganoderma*, named lingzhi in China, has been used for centuries as drug and nutraceutical to treat diseases. Based on our research and other literatures, the chapter summarizes the progress of preparation, structural features and properties, bioactivities of *Ganoderma* polysaccharides. The aim is to provide a comprehensive source of information for researchers and consumers of *Ganoderma*, so they can better understand *Ganoderma* polysaccharides and their biological activities. In addition, more clinical studies should be carried out to meet the criteria for new drug development, and more convincing scientific data should be provided. In addition, on the basis of a large number of studies on *Ganoderma* polysaccharides, we suggest that more clinical studies should be carried out so that *Ganoderma* can be better recognized and applied all over the world.

Keywords *Ganoderma* · Lingzhi · Polysaccharide · Structure · Bioactivity

4.1 Introduction

Ganoderma, named lingzhi in China, has been used for centuries as drug and nutraceutical to treat assorted diseases, improve health, and prolong life in China. It is a medicinal fungus which has been studied systemically during the past 40 years [1]. Over 400 different biologically active constituents were found in fruiting bodies, mycelia, and spores, principally polysaccharides, triterpenoids, proteins, enzymes, steroids, sterols, nucleotides, fatty acids, vitamins, and minerals. A large number of literatures have proved *Ganoderma* to have different therapeutical properties to control various diseases. Broad spectra of its pharmacological actions have been established, which include immunomodulation, anticancer, antidiabetic, antioxidant, antiatherosclerotic, antifibrotic, chemopreventive, antitumor, anticancer drug toxicity prevention, analgesic, antiinflammatory, antinociceptive, antimicrobial, hypolipidemic, hepatoprotective, antiandrogenic, antiangiogenic, antiherpetic, antiarthritic,

J. Zhang (✉) · Y. Liu · Q. Tang · S. Zhou · J. Feng · H. Chen
Institute of Edible Fungi, Shanghai Academy of Agricultural Sciences, Shanghai, China

Z. Lin, B. Yang (eds.), *Ganoderma and Health*, Advances in Experimental Medicine and Biology 1181, https://doi.org/10.1007/978-981-13-9867-4_4

antiosteoporotic, antiaging, antiulcer properties, and estrogenic activities. Polysaccharides and triterpenoids are considered as the main active compounds in *Ganoderma* [2].

Present pharmacological and clinical investigations have confirmed that *Ganoderma* exhibits significant antitumor effect, which is generally achieved by boosting host immune function. The immune function potentiation mediated by *Ganoderma* polysaccharide is considered to be a core mechanism of antitumor action. Besides immunomodulation activity, the polysaccharides were also found to have anticancer, antidiabetic, antioxidant, antiinflammatory, and antiaging activities [3]. *Ganoderma* polysaccharide has always been a hot study spot in medicinal mushroom research. *Ganoderma* polysaccharides have been developed into medicines, healthy foods, and dietary supplements and are popular all over the world. It is estimated that sales of *Ganoderma* products have exceeded $4 billion [4]. This chapter aims to investigate and summarize the preparation, structural elucidation, bioactivity studies, and clinical trials of *Ganoderma* polysaccharides.

4.2 Preparation of *Ganoderma* Polysaccharides

In *Ganoderma*, the content of polysaccharide takes about 10% of the dry weight. Obtaining the polysaccharides requires a series of preparation procedures, including material selection, pretreatment, extraction, and purification. Besides of some traditional technologies of extraction and purification used for many years, several modern equipment and technologies have also been introduced in recent researches of *Ganoderma*.

4.2.1 Material Selection for Ganoderma *Polysaccharide Extraction*

Ganoderma polysaccharide can be extracted from either fruiting bodies or spores from cultivation or mycelia from fermentation broth. The content of polysaccharide varies with the source, strain, culture medium, and harvesting time. It was found that water-soluble polysaccharide content in fruiting bodies was about 1.6%. And the content in mycelia was about 6 times higher [5]. The polysaccharide content in spores was about 1.5–3% [6, 7]. The β-D-glucan content in fruiting bodies could range from 0.01% to 0.8% in materials acquired by different strains or cultivation patterns [8]. During the fruiting cycle of *Ganoderma*, it is found that primordium and young fruiting bodies contained higher content of soluble polysaccharides than that from the mature fruiting bodies, especially for the fraction with million molecular weight [9]. The solubility of polysaccharides in *Ganoderma* decreases when the fruiting body matures, which increases the difficulty of extraction.

4.2.2 *Pretreatment on* Ganoderma *Before Polysaccharide Extraction*

After selecting out the suitable material, pretreatment on the material of *Ganoderma* is required, because this procedure can increase the polysaccharide yield and reduce impurities in extracts, thus reducing the difficulty of the purification step. After pretreatment by 95% ethanol extraction, low molecular weight and liposoluble constituents such as triterpenes, sterols, and lipids can be removed from the material. Carbohydrates with molecular weight less than 1000 Da, including monosaccharides, oligosaccharides, and polyol could be extracted out by 80% ethanol [10, 11]. These substances take about up to 20% of the dry weight of extract. After pretreatment by ethanol, the content of polysaccharide in water extract will increase significantly.

Apart from impurity removal pretreatment, grinding material is also a useful pretreatment step, especially for the fruiting bodies of *Ganoderma*. In recent researches, grind technologies such as superfine grinding and nano-grinding were found to reduce the granularity and increase the polysaccharide yield by three to five times [12]. In addition, pretreatment by enzymes with different functions, such as papain, pectinase, trypsin, or cellulose, is helpful to remove impurities. Some enzymes can partly degrade the cell wall of the material, which are composed of chitin, glucan, and protein. By these assistance steps, the polysaccharide extracting yield will be significantly increased [13, 14].

4.2.3 *Extraction of* Ganoderma *Polysaccharide*

Polysaccharides in *Ganoderma* are soluble in water, weak acid or weak base, but not soluble in organic solvents such as alcohol, ether, and acetone. Because of this feature, hot water extraction is the most widely used extracting method nowadays. Commonly the *Ganoderma* material is extracted by boiling water, because high temperature will expand the material greatly, help the release of intracellular substances, and increase the solubility of polysaccharides. In conventional experiments, the *Ganoderma* material is extracted with water at 95~100 °C at the material-liquid ratio from 1:10 to 1:15 for 1 h, and repeated two or three times. To extract polysaccharides with molecular weight of millions or insoluble in neutral water, some chemical agents such as alkaline or acid at concentration of 0.1~1.0 mol/L are used. It was found that using 5.1% NaOH solvent at 60.1 °C for 77.3 min, the polysaccharide yield increased 2 time folds [15]. There are also several new extraction technologies for extracting polysaccharide of *Ganoderma*, such as ultrasonic, microwave, or high pressure. Polysaccharide extracted with ultrasonic possessing has relatively lower molecular weight and higher total content than extracted with hot water [16, 17]. Higher β-glucan yield could be achieved by using pressurized liquid extraction than microwave-assisted extraction [18].

4.2.4 *Separation and Purification of* Ganoderma *Polysaccharide*

4.2.4.1 Impurity Separation

In the extraction procedure, not only polysaccharides but also a lot of proteins are extracted. To separate out proteins, savage reagent (chloroform: *n*-butanol = 5:1) is usually used to reduce the free protein in the extracts [19]. Protease, TCA precipitation, and $CaCl_2$ salting out were also used in recent studies, aiming to remove proteins from extracts at lower costs [20].

Macroporous resin was also tested to remove proteins, pigments, and other impurities from polysaccharide extracts in the past few years. Using static or dynamic absorption, the decoloration rate can reach 90%. The protein removal rate and polysaccharide retention rate can reach 70% [21, 22].

Three-phase extraction technology is a new separation method which has been used in polysaccharide separation of medicinal mushrooms [23, 24]. It is a simple, efficient, and green bio-separation technology. This method normally uses ammonium sulfate and organic solvent to remove organic impurity from crude extracts, such as proteins and pigments. Organic solvents including isopropanol, ethyl acetate, and *t*-butanol can denature the protein in extracts, and the denatured protein will then be precipitated out of aqueous phase due to the salting out effect. Temperature and pH value are the main factors affecting the polysaccharide yield in three phase extraction.

4.2.4.2 Fractionation

After impurity was removed, the extract usually contains several different polysaccharide fractions. As polysaccharides with different molecular weights possess different solubility in ethanol or methanol solvents, polysaccharides in the extracts can be fractionated successively using ethanol precipitation with different percentages. Typical ethanol percentages used in practice are 30%, 60%, and 90%. Polysaccharides with the highest molecular weight will be precipitated in the 30% fraction firstly. The precipitates can be collected by centrifuge and washing with the same percentage of ethanol solvent for two to three times.

Ultrafiltration technology can be used to separate polysaccharides based on the molecular weight using ultrafiltration membrane under certain transmembrane pressures. Based on membranes with different molecular weight cutoffs, the polysaccharides in extract are separated into fractions according to molecular weight. There are several types of membrane used in ultrafiltration, including coil membrane and hollow-fiber membrane. Ultrafiltration method is a faster and more suitable method than ethanol precipitation for large-scale production.

4.2.4.3 Purification by Chromatography

After separation, chromatography will be used to obtain a purified polysaccharide. Ion-exchange chromatography, gel-exclusion chromatography, and affinity chromatography are the main chromatography methods.

Ion-exchange chromatography uses cellulose, sephadex, or sepharose as stationary phase, binding with diethylaminoethyl (DEAE) anionic group. NaCl solution or phosphate buffer are often used as mobile phases. After the samples are loaded on filler in column, the column is eluted by isocratic elution, stepwise elution, or gradient elution, the polysaccharide absorbed weakly is eluted firstly. This method is popularly used in polysaccharide purification because of its moderate absorption and elution. After dialysis of collected eluent, concentration, and frozen drying, the preliminary purified polysaccharide fractions were obtained.

Size-exclusion gel chromatography is a method to separate fractions according to the size of molecular weight. Sepharose, sephadex, and Bio-Gel P are ordinarily used as stationary phases for polysaccharide purification. There are several gel patterns suitable for purification of different molecular weight samples. Water or buffers with different ionic strengths are used as eluents. Adding small amount of bacteriostatic agent in eluent, such as sodium azide or 4-chlorobutanol are also necessary for long-term use. Polysaccharide fractions will be eluted out successively from high to low molecular weight.

Affinity chromatography uses sepharose with protein, trysyl, heparin, or boronate as stationary phase. Concanavalin and boronate are often used for polysaccharide purification [25, 26]. NaCl solution or phosphate buffer can be used as absorption mobile phase, and sorbitol, phosphate buffer, or boric acid can be used as elution mobile phase [27, 28]. The pH value or concentration of eluent would affect the absorption and elution of polysaccharides.

4.3 Structural Features and Properties of *Ganoderma* Polysaccharides

Recent studies reported that more than 200 polysaccharide fractions have been isolated and purified from the fruiting bodies, spores, mycelia, and cultivation broth of *Ganoderma* [29]. Their chemical structures and properties, such as monosaccharide compositions, glycosidic bond linkage types, branch characteristics, molecular weights, and chain conformations, have been investigated by chemical analysis, spectrum analysis, and chromatography technology. Research results showed that most polysaccharides from *Ganoderma* were heteropolysaccharides with different combinations of glucose, mannose, galactose, xylose, fucose, and arabinose, and (1→3)-, (1→4)-, (1→6)-α, and β-glucans were also identified. Comparing with other biopolymers such as proteins and nucleic acids, polysaccharides had the greatest potential for structural variability [30], which results in the difficulties for establishing the structure-activity relationship for *Ganoderma* polysaccharides.

4.3.1 Available Techniques for Structure Elucidation

In order to elucidate the complex structure of polysaccharides, some advanced technologies, such as HPSEC-MALLS-RI, HPAEC, GC–MS, FT-IR, and NMR, are required to be used together with chemical reactions as methylation analysis, partial hydrolysis, periodate oxidation, and controlled Smith degradation. To understand the primary structure of a polysaccharide, the information of molecular weight distributions, monosaccharide compositions with the corresponding molar ratios, linkage positions between the glycosidic linkages and branches, distinction of furanosidic and pyranosidic rings, α- or β-configuration of the glycosidic linkages, sequences of monosaccharide residues and repeating units, and positions of some substitutions (including O-phosphorylation, O-acetylation, O-sulfation, etc.) need to be known.

4.3.1.1 Purity Analysis and Molecular Weight Determination

The first step of characterizing a polysaccharide is to determine its purity, which is reflected by its chemical composition, such as total sugar content and levels of uronic acid and proteins. Colorimetric methods are suitable for estimating the contents of these factors. High-performance size-exclusion chromatography (HPSEC) coupled with suitable columns and RI detector are normally used to determine the homogeneity and average molecular weight (M_w) of polysaccharides. Homogeneous polysaccharides commonly showed single symmetrical peaks in HPSEC analysis. The corresponding molecular weights are calculated by high-performance gel permeation chromatography (HPGPC) based on a series of standards with different molecular weights. Recently, HPSEC systems equipped with multiple-angle laser light scattering detectors (MALLS), UV, and refractive index detectors (RI) are used to determine the M_w, shape and conformation of polysaccharides, as MALLS is convenient for providing precise information on the molecular weight of the polysaccharide with no standards.

4.3.1.2 Monosaccharide Composition Analysis

The second step is to determine the monosaccharide compositions and the corresponding molar ratios of different monosaccharides, which are usually determined by high-performance liquid chromatography (HPLC) or gas-liquid chromatography (GC) after a complete acid hydrolysis. For GC analysis, a previous derivatization step is necessary, and different methods of derivatization have been developed to meet the requirements of analysis. In order to analyze neutral, amino, and acidic monosaccharides as their alditol acetates, Sassaki et al. [31] described a simple modification step for acetylation, which requires small amounts of sample and solvents [31]. High-performance anion-exchange chromatography coupled to pulsed amperometric detection (HPAEC-PAD) is a powerful analytical technique with high-resolution separation and sensitive quantification for monosaccharides, so it is widely used in monosaccharide analysis due to its convenience without any previous derivatization.

4.3.1.3 Methylation Analysis

The glycosidic linkage position of sugar residues in a polysaccharide is established by methylation analysis, which is the most frequently used technique to determine the linkage types of residues. The derivatization of a polysaccharide mainly includes the process of methylation, acid hydrolysis, reduction, and acetylation. The resultant partially methylated alditol acetates (PMAA) are obtained and analyzed by a GC–MS system. The individual peaks of the PMAA and fragmentation patterns are identified by their mass spectra and relative retention time in GC. The percentages of methylated sugars are estimated as ratios of the peak areas. These results are used to give a precise analysis of the linkages, proportions of branching points, and non-reducing end units.

4.3.1.4 Nuclear Magnetic Resonance (NMR) Spectroscopy Analysis

Based on the linkage type information obtained from methylation analysis, NMR is a very useful tool to provide detailed structural information of polysaccharides, including α- or β-anomeric configurations, linkage patterns, and sequences of the repeating units in polysaccharides. Polysaccharides need to be dissolved in deuterium solvents or the chosen solvent (around 20–50 mg/mL). According to the NMR, α or β configuration of residue is easily determined by chemical shifts of ^{1}H or ^{13}C of polysaccharide in anomeric regions. For example, most of the α-anomeric protons will appear in the region of 5–6 ppm, and the corresponding carbon signals mostly appear in the region of 95–103 ppm. The β-anomeric protons and carbons will appear in the 4–5 ppm region and in the 101–105 ppm region, respectively. The chemical shifts of different protons and carbons in polysaccharides are assigned in Fig. 4.1a, b.

1D ^{1}H and ^{13}C NMR spectrum normally contain all the structural information of the polysaccharides. However, the signals are frequently crowded in narrow regions. Hence, the interpretation of 1D NMR spectra becomes difficult if a polysaccharide contains many similar residues. The development of two- and multi-dimensional NMR techniques has significantly improved the resolution and sensitivity of NMR spectroscopy. For example, the homonuclear correlated spectrum COSY will overcome most of the overlap problems in the 1D spectrum; with the assistance of

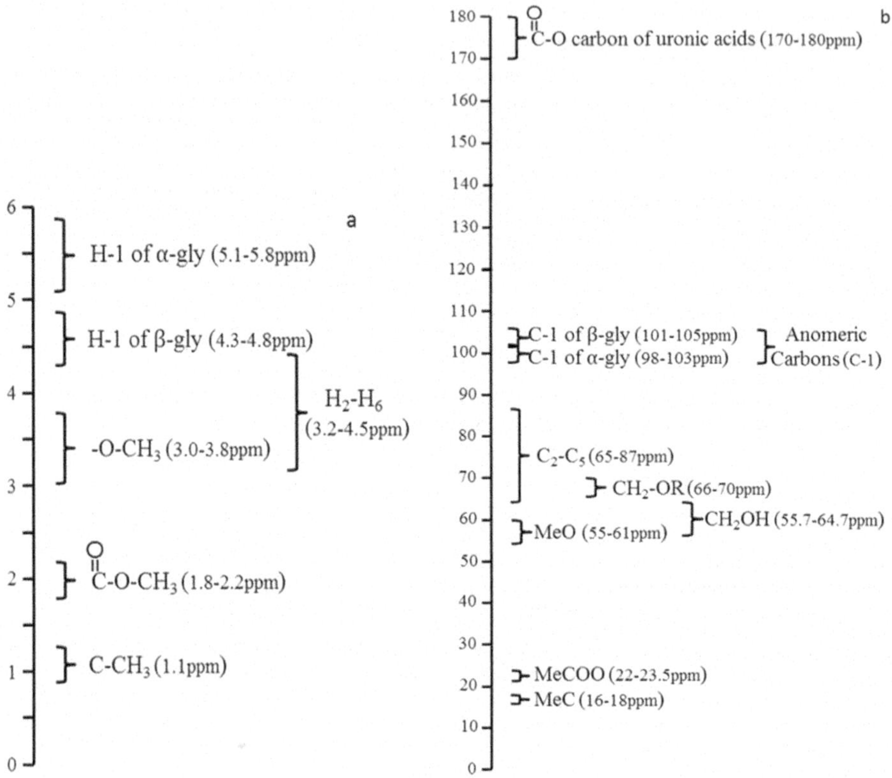

Fig. 4.1 Illustration of chemical shifts of polysaccharides in ^{1}H (**a**) and ^{13}C (**b**) NMR spectroscopy. (Adapted from Cui and Wang [32])

TOCSY, the identification of sugar residues is extremely useful for assigning ^{1}H resonances. The complete assignment of ^{13}C-resonances is achieved by ^{1}H-^{13}C heteronuclear correlation. Long-range correlation techniques, such as nuclear overhauser enhancement (NOE) and heteronuclear multiple bond correlation (HMBC), are most useful to provide sequence information of polysaccharides. Of course, there are many other powerful 2D NMR techniques for structural analysis of complex polysaccharides, and new techniques are still being developed. For further details in the field, please follow the suggested references and pay attention to publications in the current literature [32, 33].

4.3.2 *Structural Features of Polysaccharides Purified from* Ganoderma *Species*

With the development of modern techniques, hundreds of polysaccharide fractions have been isolated and purified from different materials of *Ganoderma*, and the primary structural features of some fractions were elucidated. These research results have provided more information for further understanding the structure-function relationship of polysaccharides. The representing structures including glucans, heteroglucans, and heterogalactans and heterogeneous main-chain polysaccharides are summarized as follows.

4.3.2.1 Glucans

Glucans are structurally variable polymers of D-glucopyranose (D-Glc*p*) [34]. Despite of simplicity of their glucose monosaccharide composition, large diversity can be found concerning number and anomeric configuration of D-Glc*p* units, position, and sequence of glycosidic bonds along a chain, branching degree and chain conformation. β-D-glucans with (1→3), (1→4) and (1→6) linkages are the main glucans found in *Ganoderma* materials. It is generally known that linear (1→3)-β-D-glucans, especially those having β-D-glucopyranosyl units attached by (1→6) linkages as single-unit branches, are biological response modifiers (BRMs) as they could enhance the immune system systemically [35]. Therefore, study on fungal β-D-glucans is a hot pot in the research field of natural medicine and functional food. Some scientific reports showed that homoglucans from the water extracts of *G. lucidum* fruiting bodies are glucans with β-(1→3)-linked-D-glucopyranosyl as backbone and with β-(1 →6)-D-glucopyranosyl as branches. They may show different branching degrees and different ranges of molecular weight distributions from thousands to millions. Liu et al. [36] have reported a homoglucan (GLP20) with molecular weight of 2.42×10^6 g/mol based on the HPSEC-MALLS-RI determination, which was purified from *G. lucidum* fruiting bodies with a simple ethanol precipitation method [36]. Its structure was elucidated as a β-(1→3)-D-glucan with β-(1→6) branches (DB ~ 0.33), and its conformation was proved to be triple-helical chains in aqueous solution [37]. The structures of several glucans from *G. lucidum* spores are more complex than that from the fruiting bodies. They normally contained different combinations of glucose, including linear (1→3)-α-D-glucan [38], branched β-D-glucan with (1→4)-α-D-glucan backbone [39], β-D-glucan with (1→3)-β-D-glucan backbone [40], β-D-glucan with (1→6)-β-D-glucan backbone [7], and mixed-linkage α, β-D-glucans [41]. A water-soluble β-D-glucan (GLSWA-I) with an average molecular weight of 1.57×10^5 g/mol was isolated from the aqueous extract of the wall-breaking spore powder *of G. lucidum* and purified by an anion-exchange and gel-permeation chromatography [6]. Based on the monosaccharide composition analysis, methylation analysis, and 1D and 2D NMR

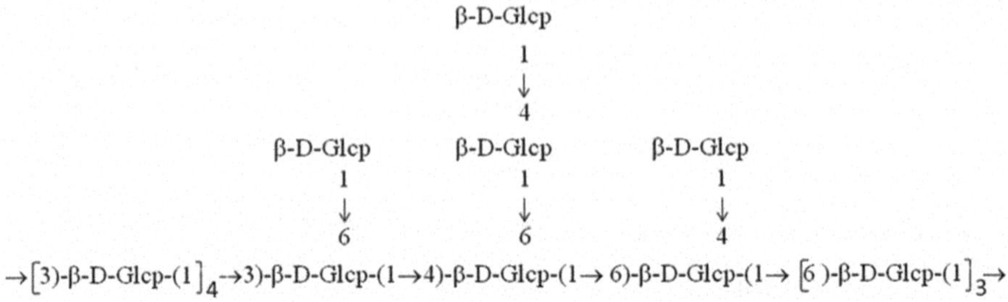

Fig. 4.2 Primary structure of polysaccharide GLSWA-I obtained from *G. lucidum* spores. (Adapted from Wang et al. [6])

spectroscopy, the repeating unit of polysaccharide GLSWA-I was elucidated as follows (Fig. 4.2).

Dong et al. [7] = of D-glucan (GLSA50-1B) from *G. lucidum* spores using monosaccharide composition analysis, methylation analysis, partial acid hydrolysis, acetolysis, and NMR and ESI-MS spectroscopy [7]. The results indicated that GLSA50-1B was a novel β-D-glucan featured by a 1,6-linked β-D-glucopyranosyl backbone with different lengths of branches consisting of terminal and 1,4-linked glucopyranosyl residues, attached to O-4 of alternative glucose residues in the backbone.

The complexity of glucans can further increase when there are monosaccharides other than glucose present or structural differences in chain conformation, branching degree, molecular weight, or presence of functional groups. All these differences result in differences in structural properties and immunological activities of glucans.

4.3.2.2 Heteroglucans

Heteroglucans, consisting of a main chain composed by D-glucopyranose residues and branches with different monosaccharides, are the most common polysaccharides present in mushrooms. The structures of heteroglucans purified from *Ganoderma* usually are composed of α or β-D-Glc*p* residues with (1→3)-, (1→4)-, and/or (1→6)-linkages, being substituted at O-2 or O-6 by terminal non-reducing residues of α or β-D-Glc*p* with small amount of rhamnose, galactose, and mannose. Disaccharides or small branches were also observed such as (1→6)-β-D-Glc*p* and (1→4)-β-D-Gal*p*, T-Glc*p*-(1→6)-Glc*p*-(1→, T-Glc*p*-(1→4)-GlcA*p*-(1→, or T-Glc*p*-(1→4)-GlcA*p*- (1→6)-Glc*p*-(1→. Some more details about the structures, such as molecular weight and purification procedures, are shown in Table 4.1.

Table 4.1 Source, fractionation method, and structural characteristics of polysaccharide fractions from *Ganoderma* spp.

Source	Sample No.	Extraction and purification method	Backbone chain	Branching positions	Side chain	Ratio of branches to main chain	Molecular weight	References
Glucan								
G. lucidum (fruiting bodies)	GLP20	Hot water extraction; precipitated with 20% ethanol	→3)-β-D-Glc*p*-(1→ →3,6)-β-D-Glc*p*-(1→	O-6	β-D-Glc*p*-(1→	1:3	3.75 × 10^6 g/mol (static laser light scattering) 2.42 × 10^6 g/mol (HPSEC-MALLS-RI)	[36, 37]
G. lucidum (fruiting bodies)	PL3	Hot water extraction; DEAE-cellulose; Sephacryl S-300 HR	→3)-β-D-Glc*p*-(1→ →3,6)-β-D-Glc*p*-(1→	O-6	β-D-Glc*p*-(1→	/	6.3 × 10^4 Da	[45]
G. lucidum (fruiting bodies)	GLG	Hot water extraction, ethanol precipitation, Toyopearl HW-55F	→3)-β-D-Glc*p*-(1→ →3,6)-β-D-Glc*p*-(1→	O-6	β-D-Glc*p*-(1→	/	1.8 × 10^6 g/mol (HPSEC-MALLS-RI)	[46]
G. lucidum (fruiting bodies)	GLPs	Extracted with 1 M NaOH at 40 °C, dissolved in DMSO and dialyzed (M_w cutoff 8000) with distilled water to get a water-insoluble white precipitate	→3)-β-D-Glc*p*-(1→	O-6 O-2	β-D-Glc*p*-(1→	3.1% 1.3%	13.3 × 10^4 g/mol (SEC-LLS, in DMSO)	[47]

(continued)

Table 4.1 (continued)

Source	Sample No.	Extraction and purification method	Backbone chain	Branching positions	Side chain	Ratio of branches to main chain	Molecular weight	References
Ganoderma sinense (fruiting bodies)	GSP-2	DEAE Sepharose CL-6B, Sephacryl S300 and S400 gel permeation chromatography	→4)-β-D-Glc*p*-(1→ →6)-β-D-Glc*p*-(1→ →3,6)-β-D-Glc*p*-(1→	O-3	β-D-Glc*p*-(1→ →3)-β-D-Glc*p*-(1→	/	3.2×10^4 Da (HPGPC)	[48]
Ganoderma resinaceum (fruiting bodies)	DESSK5	Alkaline extraction, dialysis and ethanol precipitation, ultrafiltration	→3)-β-D-Glc*p*-(1→ →3,6)-β-D-Glc*p*-(1→	O-6	β-D-Glc*p*-(1→ →4)-β-D-Glc*p*-(1→	1:2	2.6×10^4 (HPSEC-MALLS-RI)	[49]
G. lucidum (spores)	SP	Water extraction, ethanol precipitation, DEAE-cellulose, Sephacryl S-200 HR	→3)-β-D-Glc*p*-(1→ →3,6)-β-D-Glc*p*-(1→	O-6	β-D-Glc*p*-(1→ →3)-β-D-Glc*p*-(1→ →6)-β-D-Glc*p*-(1→	/	1×10^4 (HPGPC)	[40]
G. lucidum (spores)	GLSWA-I	Hot water extract residues, superfine grinding, water extraction, ethanol precipitation, DEAE Sepharose, Sephacryl S-300HR	→3)-β-D-Glc*p*-(1→ →3,6)-β-D-Glc*p*-(1→ →4)-β-D-Glc*p*-(1→ →4,6)-β-D-Glc*p*-(1→	O-6 O-4	β-D-Glc*p*-(1→ →4)-β-D-Glc*p*-(1→	/	1.57×10^5 g/mol (HPSEC-MALLS-RI)	[6]

Source	Sample No.	Extraction and purification method	Backbone chain	Branching positions	Side chain	Ratio of branches to main chain	Molecular weight	References
G. lucidum (spores)	GLSA50–1B	Hot water extraction, graded ethanol precipitation DEAE-cellulose, Sephacryl S-300	→6)-β-D-Glc*p*-(1→ →4,6)-β-D-Glc*p*-(1→	O-4	β-D-Glc*p*-(1→ →4)-β-D-Glc*p*-(1→	/	1.03×10^5 (HPGPC)	[7]
G. capense (mycelia)	GCP50-1	Hot water extraction, DEAE Sepharose CL-6B, Sephadex G-75	→4)-α-D-Glc*p*-(1→ →4,6)-α-D-Glc*p*-(1→	O-6	α-D-Glc*p*-(1→		1.5×10^4 Da (HPGPC)	[50]
Heteroglucan								
G. lucidum (fruiting bodies)	LZ-D-7	Hot water extraction, DEAE Sepharose fast flow, Sephacryl S-300 and S500	→4)-D-Glc*p*-(1→ →2,6)-D-Glc*p*-(1→	O-2 or O-6	α-D-Glc*p*-(1→ With small amount of rhamnose, galactose, mannose	/	1×10^6 Da (HPGPC)	[51]
G. lucidum (fruiting bodies)	GLP2	Hot water extraction, DEAE-cellulose-32, Sephacryl S-200 HR	→3)-D-Glc*p*-(1→ →4)-D-Glc*p*-(1→ →6)-D-Glc*p*-(1→ →4,6)-D-Glc*p*-(1→ →3,6)-D-Glc*p*-(1→	/	D-Glc*p*-(1→ Little amount of galactose and mannose	/	1.54×10^4	[52]
G. lucidum (fruiting bodies	GLP-F1-1	Ultrasonic/microwave-assisted extraction, DEAE Sepharose Fast Flow, Sephacryl S-500HR	→4)-β-D-Glc*p*-(1→ →4,6)-β-D-Glc*p*-(1→	O-6	β-D-Glc*p*-(1→ →6)-β-D-Glc*p*-(1→ →4)-β-D-Gal*p*-(1→		2.5×10^6 Da (HPGPC)	[53]

(continued)

Table 4.1 (continued)

Source	Sample No.	Extraction and purification method	Backbone chain	Branching positions	Side chain	Ratio of branches to main chain	Molecular weight	References
Ganoderma atrum (fruiting bodies	PSG-1-$F_{0.2}$	Hot water extraction, ethanol precipitation, Superdex-200, Q-Sepharose Fast Flow	→3)-β-D-Glc*p*-(1→ →3,6)-β-D-Glc*p*-(1→	O-6	T-Glc*p*-(1→6)-Glc*p*-(1→ T-Glc*p*-(1→4)-GlcA*p*-(1→ T-Glc*p*-(1→4)-GlcA*p*-(1→6)-Glc*p*-(1→	/	/	[54]
Heterogalactan								
G. lucidum (fruiting bodies)	LZ-D-1	Hot water extraction, DEAE Sepharose Fast Flow, Sephacryl S-300	→6)-α-D-Gal*p*-(1→ →2,6)-α-D-Gal*p*-(1→	O-2	α-L-Fuc*p*-(1→	1:5	2.8×10^4 Da (HPGPC)	[55]
Ganoderma atrum (fruiting bodies)	PSG-2	Hot water extraction, ethanol precipitation, Superdex-200	→6)-α-D-Gal*p*-(1→ →2,6)-α-D-Gal*p*-(1→	O-2	α-L-Fuc*p*-(1→ α-D-Gal*p*-(1→ α-D-Man*p*-(1→ T-α-D-Gal*p*-(1→6)-2-OAc-α-D-Gal*p*-(1→ T-α-D-Gal*p*-(1→6)--α-D-Gal*p*-(1→4)-β-D-Glc*p*-(1→	/	6.9×10^4 Da (HPGPC)	[56]

Source	Sample No.	Extraction and purification method	Backbone chain	Branching positions	Side chain	Ratio of branches to main chain	Molecular weight	References
G. lucidum (extracellular polysaccharide)	GLP-2	Fermented broth, DEAE-Sephacel, Sephadex G200	→4)-α-D-Gal*p*-(1→	O-6	L-Ara-(1→ L-Rha-(1→ →6)-α-D-Man*p*-(1→ →4)-α-D-Glc*p*-(1→	/	2.2×10^4 Da (HPGPC)	[42]
Heterogeneous main chain								
G. lucidum (fruiting bodies)	GLPCW-II	Hot water extraction, DEAE Sepharose Fast Flow, Sephacryl S-300 HR	→6)-α-D-Gal*p*-(1→ →3)-α-D-Glc*p*-(1→	O-2 of →6)-α-D-Gal*p*-(1→	α-L-Fuc*p*-(1→	/	1.2×10^4 Da (HPGPC)	[43]
G. lucidum (fruiting bodies)	LZ-B-1	Hot water extraction, DEAE Sepharose Fast Flow, Sepharose CL-6B	→6)-α-D-Gal*p*-(1→ →2,6)-α-D-Gal*p*-(1→ →3)-β-D-Glc*p*-(1→ →4,6)-β-D-Glc*p*-(1→	O-2 of →2,6)-α-D-Gal*p*-(1→ O-6 of →4,6)-β-D-Glc*p*-(1→	β-D-Glc*p*-(1→ α-L-Fuc*p*-(1→	/	1.12×10^4 Da (HPGPC)	[57]

(continued)

Table 4.1 (continued)

Source	Sample No.	Extraction and purification method	Backbone chain	Branching positions	Side chain	Ratio of branches to main chain	Molecular weight	References
G. lucidum (fruiting bodies)	LZ-C-1	Hot water extraction, DEAE Sepharose Fast Flow, Sephacryl S-300 HR	→6)-α-D-Gal*p*-(1→ →2,6)-α-D-Gal*p*-(1→ →3)-β-D-Glc*p*-(1→ →4,6)-β-D-Glc*p*-(1→	O-2 of →2,6)-α-D-Gal*p*-(1→ O-4 of →4,6)-β-D-Glc*p*-(1→	β-D-Glc*p*-(1→ α-L-Fuc*p*-(1→	/	7×10^3 Da (HPGPC)	[58]
G. lucidum (fruiting bodies)	FYGL-1	Alkaline extraction, DEAE-52, Sephadex G75	→2)-β-L-Rha*p*-(1→ →3,6)-α-D-Gal*p*-(1→ →2,6)-α-D-Glc*p*-(1→	O-3 O-2	α-D-Glc*p*-(1→		7.8×10^4 Da (HPGPC)	[44]
G. lucidum (fruiting bodies)	PL-1	Hot water extraction, ethanol precipitation, DEAE cellulose, Sephacryl S-300 HR	→4)-α-D-Glc*p*-(1→ →6)-β-D-Gal*p*-(1→ →4,6)-β-D-Glc*p*-(1→	O-6 of glucose residues and O-2 of galactose residues	Terminal glucose, 1,6-linked glucosyl residues and terminal rhamnose		8.3×10^3 Da (HPGPC)	[45]

/: no information provided

4.3.2.3 Heterogalactans

Besides glucan, several articles have reported the isolation and structural characterization of heterogalactans from *G. lucidum* and *G. atrum* fruiting bodies, as well as the mycelia of *G. lucidum*. Heterogalactans usually present a common structure with a backbone of (1→6)-linked-α-D-galactopyranosyl residues, some of which are substituted at O-2 by α-L-Fucp-(1→, α-D-Galp-(1→, α-D-Manp-(1→, or oligosaccharides of T-α-D-Galp-(1→6)-2-OAc-α-D-Galp-(1→, T-α-D-Galp-(1→6)-α-D-Galp-(1→4)-β-D-Glcp-(1→ as side chains. Li et al. [42] reported an extracellular polysaccharide (GLP-2) from *G. lucidum* cultivation broth with the (1→4)-linked-α-D-galactopyranosyl residues as the main chain and L-Ara-(1→, L-Rha-(1→, →6)-α-D-Man*p*-(1→, →4)-α-D-Glc*p*-(1→ as side chains substituted at O-6 position. More details about the heterogalactan structures, such as molecular weight and purification procedures are shown in Table 4.1.

4.3.2.4 Heteroglycan with Heterogeneous Main Chain

There are several articles reported heteroglycans with heterogeneous main chain, which were purified from *G. lucidum* fruiting bodies (in Table 4.1). The backbones of polysaccharides mainly include monosaccharides such as β-L-Rha*p*, α-D-Gal*p*, β-D-Gal*p*, α-D-Glc*p*, and β-D-Glc*p* in different linkage combinations. One fraction (GLPCW-II) presented only (1→6)-linked-α-D-Gal*p* and (1→3)-linked-α-D-Glc*p* in the main chain [43], while other fractions were described as structures with (1→6)-linked-α-D-Gal*p*, (1→3)-linked-β-D-Glc*p*, and (1→4)-linked-β-D-Glc*p* as main chains. Pan et al. [44] also reported a polysaccharide FYGL-1 containing a main chain of three different monosaccharides, including β-L-Rha*p*-(1→, →6)-α-D-Gal*p*-(1→, and →6)-α-D-Glc*p* -(1→. More detailed information about the structures is listed in Table 4.1.

In a word, polysaccharides from *Ganoderma* materials represent a structurally diverse class of biological macromolecules with a wide range of physicochemical properties. As a common medicinal and nutritional mushroom, *Ganoderma* have been widely used for the prevention or treatment of various types of diseases. Polysaccharides are found to be one of the most important components responsible for bioactivities of immunostimulatory and antitumor activities. To date, the detailed primary repeating units of some polysaccharide fractions have been elucidated. However, the higher levels of structures of these polysaccharides such as chain conformations in solution and 3D structure are still unclear. As polysaccharides are complex biomacromolecules, their structural characteristics like specific glycosidic linkages, monosaccharide composition, molecular weight, and chain conformation will affect their functions. In order to establish the structure-function relationship, the primary structures, conformational properties, and bioactivities of polysaccharides from *Ganoderma* should be further studied.

4.4 Bioactivities of *Ganoderma* Polysaccharide

Polysaccharide is one of the most important active ingredients in *G. lucidum*, which has been extensively used as medicine and functional food. The *Ganoderma* polysaccharides have been studied for several years and demonstrated to possess diverse bioactivities such as immunomodulatory, antitumor, hepatoprotective, and antihypertensive activities and so on.

4.4.1 Immunomodulatory Activity

More and more works have demonstrated that the polysaccharides from *G. lucidum* have immunomodulatory activity. It could enhance the function of lymphocytes, macrophages, dendritic cells, natural killer cell, etc. Continuous studies on the effect of polysaccharides on lymphocytes and macrophages have been done by Prof Zhang's group [59]. A bioactive fraction (named GLIS) purified from the fruiting body of *G. lucidum* stimulated the proliferation of mouse spleen lymphocytes. Most of the activated lymphocytes were B cells, which were proved by detecting the expressed CD71 and CD25 on the cell surface. GLIS was also found to increase the secretion of immunoglobulin of B cells. In addition, it was found that GLIS activated B cells through the PKC pathway [59]. After treated by GLIS, RAW264.7 macrophages were enlarged and formed pseudopodia. Treating RAW264.7 macrophages by GLIS resulted in significant increase of NO production, induction of cellular respiratory burst activity, and increased gene expression levels of IL-1β, IL-12p35, and IL-12p40 [60]. Bone marrow-derived macrophages from tumor-bearing mice also became activated after exposure to GLIS, and they produced IL-1β, TNF-α, and reactive nitrogen intermediates [61]. GLPss58, a sulfated polysaccharide from *G. lucidum*, inhibited the activation of complement systems and blocked the binding of TNF-α and IFN-γ to their antibodies, which inhibited not only the L-selectin-mediated inflammation, but also the complement system and cytokine-mediated inflammation [62].

Similar reports also found that the polysaccharides could activate the lymphocytes and macrophages. *G. lucidum* polysaccharides (GLPS) activated B cells and macrophages but not T cells of BALB/c mouse in vitro and produced IL-1β by peritoneal macrophages [63]. Water-soluble extracts F3 of *G. lucidum* stimulated cell proliferation and cytokine expression in mice spleen. The populations of $CD^{14+}CD^{26+}$ monocyte/macrophage, $CD^{83+}CD^{1a+}$ dendritic cells, and $CD^{16+}CD^{56+}$ NK cells were 2.9, 2.3, and 1.5 times, respectively, higher than those of the untreated controls ($p < 0.05$) [64]. The polysaccharide fractions from *G. lucidum* activated human peripheral blood mononuclear cells and induced innate inflammatory cytokines tumor necrosis factor alpha (TNF-α), interleukin (IL)-12, and interferon gamma (IFN-γ) [25].

The dendritic cells also could be activated by the polysaccharide. Polysaccharides from *G. atrum* (PSG-1) induced the maturation of dendritic cells; the molecule expression of MHC-II, CD80, and CD86; and the production of IL-12 p70, IL-6, IL-10, RANTESMIP-1α, and MCP-1 in DCs. PSG-1 directly induced DCs maturation via activating MAPK 18 pathways and indirectly stimulated DCs separated by intestinal epithelial cells [65]. The polysaccharide fractions from *G. lucidum* (Gl-PS) promoted the maturation of cultured murine bone marrow derived DC in vitro and the immune response initiation induced by DC [66].

The immune activities were also tested in animals. The polysaccharides from *Ganoderma* GPs were fed to Wistar rats, and it had immune effects in increasing the degree of toe swelling and enhancing the primary immune response to SRBC ($P < 0.01$) [67]. A polysaccharide fraction PS-F2 stimulated OVA-specific antibody production and primed IFN-γ production in OVA-specific T lymphocytes. PS-F2-adjuvated immunization also induced OVA-specific CTLs, which protected mice from a challenge of tumor cells expressing OVA [68]. A polysaccharide from the mycelia of *G. lucidum* significantly upregulated the expression of occludin, nuclear factor-κB p65, and secretory immunoglobulin A in ileum; markedly improved the levels of IFN-γ, IL-2, and IL-4; and decreased the level of diamine oxidase in serum [69]. IgA secretion from Peyer's patch cells and alpha-defensin-5 (RD-5) and RD-6 expression in the rat small intestine were increased [70]. The polysaccharides of *G. lucidum* extracted by ultrasonic/microwave (UMP) improved the weight of immune organ of immunocompromised mice, restored delayed-type hypersensitivity (DTH) reaction to DFNB, and improved hemolysis antibody level and natural killer cell activity at high dose [71]. A water-soluble extract from *G. lucidum* mycelia (MAK) has anti-inflammatory effect in murine colitis induced by trinitrobenzene sulfonic acid. Induction of granulocyte macrophage colony-stimulating factor (GM-CSF) by MAK may result in its anti-inflammatory effects [72].

4.4.2 Antitumor Activity

Polysaccharides could not inhibit tumor cells directly; it exert the antitumor effect through enhancing the host's defense system [72]. GLIS markedly increased phagocytosis of macrophage, which raised the macrophage-mediated tumor cytotoxicity. Treating mice with GLIS caused an inhibition of mouse sarcoma S180 tumor growth by 60% in vivo [61]. *G. lucidum* polysaccharide fractions GP-1 and GP-2 increased the proliferation and pinocytic activity of macrophages significantly and showed an inhibitory effect on cancer cells [73]. *G. lucidum* polysaccharide (GLP-Au) significantly promoted the proliferation of CD^{4+} and CD^{8+} T cells in splenocytes. DC/T cell co-culture study proved that GLP-Au activation on DC directly resulted in T cell proliferation. GLP-Au exhibited strong inhibitory effects on tumor growth and pulmonary metastasis when combined with doxorubicin [74].

It is found that *G. lucidum* polysaccharide could inhibit tumor growth in vivo. Treating mice with a *G. lucidum* polysaccharide significantly reduced the tumor weight [75]. The *G. lucidum* polysaccharide (GLP) also exerted antitumor effect on solid tumor induced by Ehrlich's ascites carcinoma cells. GLP (100 mg/kg) leads to 80.8% and 77.6% reduction in tumor volume and tumor mass, respectively, when administered after tumor cell implantation. Furthermore, GLP at the same dose, when administered prior to tumor inoculation, showed 79.5% and 81.2% inhibition of tumor volume and tumor mass, respectively [76].

Studies on the mechanism of *G. lucidum* polysaccharides in inhibiting tumors were also investigated. PSG-1 did not kill CT26 cells directly, but inhibited the proliferation of CT26 cells via the activation of peritoneal macrophages (MΦ). In vivo, PSG-1 significantly suppressed the tumor growth in CT26 tumor-bearing mice. PSG-1 could activate macrophages via TLR4-dependent signaling pathways, improve immunity, and inhibit tumor growth [77]. Through comparing the inhibition effects of CSGAP against Sarcoma 180 in vivo by the polysaccharides isolated from fruiting bodies (FGAP) and submerged fermentation system (SGAP), the results suggested that the carboxylate groups played a major role in antitumor activity of *G. applanatum* polysaccharide [78]. *G. atrum* polysaccharide (PSG-1) inhibited the tumor growth, possibly in part by enhancing the induction of apoptosis through cAMP-PKA signaling pathway and downregulation of Ca^{2+}/PKC signal pathway, which activates host immune function in S180-bearing mice [79].

4.4.3 Hypoglycemic Activities

The polysaccharides from *Ganoderma* had the effects of hypoglycemic in vivo, which could enhance the insulin sensitivity or increase insulin levels. Insulin and *G. lucidum* polysaccharides were administered orally to the T2DM rats induced by high-fat diet (HFD) and streptozotocin (STZ); it was found that the polysaccharides significantly improved the glucose and lipid metabolism-related parameters, which seemed to be related to enhancing the insulin sensitivity, increasing the glycogen synthesis, and facilitating the glucose transportation by regulating the PI3K/Akt pathway [80]. A neutral hyperbranched proteoglycan ingredient FYGL extracted from *G. lucidum* enhanced insulin-stimulated glycogen synthesis in HepG2 cells and lowered blood glucose in insulin resistance model mice [81]. *G. lucidum* polysaccharides also have been reported to have hypoglycemic activity by increasing plasma insulin levels and decreasing plasma sugar levels in mice [82]. A bioactive β-heteropolysaccharide fraction in *Ganoderma* may be associated with downregulation of the hepatic glucose-regulating enzyme mRNA levels via AMPK activation, improvement of insulin resistance, and decrease of epididymal fat/BW ratio [83].

The hypoglycemic activity of polysaccharide from *G. lucidum* was used in patients with type II diabetes mellitus and was demonstrated to be efficacious and safe in lowering blood glucose concentration [84]. The possible mechanisms was to reverse alloxan-induced islet viability loss by inhibiting the free radicals production, increasing serum insulin, and reducing serum glucose levels [85].

4.4.4 Hepatoprotective Effect

G. lucidum polysaccharides GLPS (0.1, 0.3, and 0.6 mg/ml) were added to the primary hepatocytes before, after, and both before and after the incubation of the hepatocytes with CCl_4 at the concentration of 8 mM in the culture medium. GLPS significantly improved cell viability and inhibited the elevations of the marker enzymes (GOT, GPT, LDH) and MDA induced by CCl4 and markedly increased the level of SOD. Treatments with GLPS resulted in a significant decrease in the expressions of CYP1A and CYP3A and significantly downregulated extrinsic apoptosis and immune inflammatory response [86]. *G. lucidum* spore has effect on [Cd(II)]-induced hepatotoxicity in mice; it protected against Cd(II)-induced liver injury in a dose-dependent manner; decreased Cd(II) accumulation in hepatic nuclei, mitochondria, and microsomes; and induced hepatic metallothionein-1 mRNA eightfold [87]. *G. atrum* polysaccharides (PSG) presented a practical option for the management of drug-induced liver injury [88].

These hepatoprotective effects may be related in part to the antioxidant properties of polysaccharides. *G. lucidum* polysaccharide (GLP) inhibited iNOS protein expression in the BCG-immune hepatic damage model, which indicated that NO participated in immune liver injury induced by BCG infection. The mechanisms of protective roles by GLP for BCG-induced immune liver injury may attribute to influence NO production in mice [89]. Treatment with *G. lucidum* aqueous extracts (GLEs) significantly decreased serum ALT and AST levels, significantly increased SOD and CAT activities, and decreased MDA content in liver compared with the α-AMA control group [90].

4.4.5 Hyperlipidemia Effect

Animal feeding experiments showed that *G. lucidum* had hypolipidemic effects. Oral administration of the polysaccharide from *G. atrum* (PSG-1) at 200 or 400 mg/kg body weight significantly reduced fasting blood glucose and decreased the levels of serum total cholesterol, triglyceride, low-density lipoprotein cholesterol, free fatty acid, and insulin resistance [91]. After the rats were treated with *G. lucidum* polysaccharides at dosages of 200 mg/kg, 400 mg/kg, and 800 mg/kg, the serum TC, TG, HDL-C, LDL-C, GSH-Px, SOD, and LPO significantly decreased [92]. Oral administration of *G. lucidum* polysaccharide for 40 days resulted in a dose-dependent significant reduction of the levels of TC and TG [93].

4.4.6 Regulation of Intestinal Flora

Recently, the effect of polysaccharide on intestinal flora has been paid more and more attentions. Polysaccharides from the mycelia of *G. lucidum* (GLP) can improve intestinal barrier function, regulate intestinal immunity, and modulate intestinal microbiota. Different metabolites associated with the improvement of intestinal immunological function and the regulations of intestinal microbiota were also identified. The results provided a potential metabolomic mechanism of health-beneficial properties of polysaccharides from the mycelia of *G. lucidum*, which might be used as functional agents to regulate the intestinal functions [94]. The polysaccharides from *G. lucidum* strain S3 GLPS3 markedly alleviated the pancreatitis in mice through decreasing lipase, AMS, IFN-γ, and TNF-α level as well as increasing SOD and total antioxidant activity. Furthermore, GLPS3 altered the composition and diversity of intestinal microbiota, especially, decreased the relative abundance of phylum *Bacteroidetes*, and increased the relative abundance of phylum *Firmicutes*. At the genus level, supplementation of GLPS3 increased the relative abundance of the beneficial bacteria such as *Lactobacillales*, *Roseburia*, and *Lachnospiraceae* [95].

4.4.7 Other Bioactivities

The polysaccharide fraction GLPS from *G. lucidum* exhibited meaningful neuroprotective effects against spinal cord injury. After the administration of GLPS, decrease of caspase-3 activity in tissue, tumor necrosis factor-alpha levels, myeloperoxidase activity, malondialdehyde levels, and nitric oxide levels were observed [96]. *Ganoderma atrum* polysaccharide PSG-1 had cardioprotective effects via attenuating apoptosis and maintaining the structure of myocardial mitochondria. Meanwhile, PSG-1-evoked cardioprotection was associated with an increase of manganese SOD activity and decrease of caspases activities [97]. The polysaccharides from *G. lucidum* PGL played a positive protective role in the pulmonary fibrosis; its possible mechanism was to improve lung antioxidant ability [98].

Although polysaccharides have diverse pharmacological activities, they are rarely used clinically as drugs now, and most of them are health products and foods. This is related to the difficulty of the separation and purification of polysaccharides from *G. lucidum*. Normally the samples used for animal tests are crude polysaccharides, so it is difficult to decide if polysaccharides play a role or whether other impurities have the effect. For example, the antioxidant activities of PSG may ascribed to the phenolic and protein components, rather than the carbohydrates, which would be more responsible to the immunomodulatory activity [99].

Although the biological activity mechanisms of the polysaccharides from *G. lucidum* are not clear yet, enhancing immune function and improving intestinal flora are recognized to be important in the role of polysaccharide. The variety of functions may be related to two pathways, and the more research is needed to prove it.

4.5 The Future Research of *Ganoderma* Polysaccharide

Although the structure, biological activity, and mechanism of *G. lucidum* polysaccharides have been extensively and deeply studied in laboratories all over the world, the complexity of polysaccharides' structure leads to the diversity of their biological activities. So far no *G. lucidum* drugs have been developed in a real sense, so our future research should focus on the basic research of the structure-activity relationship of *G. lucidum* polysaccharides. The research will help us better understand the chemical basis and mechanism of *G. lucidum* polysaccharides to exert their biological activities. At the same time, more clinical studies should be carried out to meet the criteria for new drug development, and more convincing scientific data should be provided. These efforts will enable this millennium treasure of Chinese traditional medicine to move from myth to reality.

References

1. Bishop KS, Chi HJ, Kao YX, Glucina MP, Paterson RRM, Ferguson LR (2015) From 2000 years of *Ganoderma lucidum* to recent developments in nutraceuticals. Phytochemistry 114:56–65
2. Faruque AM (2018) *Ganoderma lucidum*: persuasive biologically active constituents and their health endorsement. Biomed Pharmacother 107:507–519
3. Cao Y, Xu X, Liu S, Huang L, Gu J (2018) *Ganoderma:* a cancer immunotherapy review. Front Pharmacol 9:1217
4. Cör D, Knez Ž, Knez HM (2018) Antitumour, antimicrobial, antioxidant and antiacetylcholinesterase effect of *Ganoderma Lucidum* terpenoids and polysaccharides: a review. Molecules 23(3):649
5. Zhang J, Jia W, Xing Z, Tang Q, Liu Y, Yang Y, Zhou C, Liu F (2004) Comparison of bioactivity of fruiting body and mycelia of *Ganoderma lucidum* and their purified fractions. Mycosystema 23(1):85–92
6. Wang Y, Liu Y, Yu H, Zhou S, Zhang Z, Wu D, Yan M, Tang Q, Zhang J (2017) Structural characterization and immuno-enhancing activity of a highly branched water-soluble β-glucan from the spores of *Ganoderma lucidum*. Carbohydr Polym 167:337–344
7. Dong Q, Wang Y, Shi L, Yao J, Li J, Ma F, Ding K (2012) A novel water-soluble β-d-glucan isolated from the spores of *Ganoderma lucidum*. Carbohydr Res 353:100–105
8. Liu Y, Tang Q, Yang Y, Zhou S, Wu D, Tang C, Zhang Z, Yan M, Feng J, Zhang JS (2017) Characterization of polysaccharides from the fruiting bodies of two species of genus Ganoderma (Agaricomycetes) and determination of water-soluble beta-D-Glucan using high-performance liquid chromatography. Int J Med Mushrooms 19(1):75–85
9. Zhou S, Tang Q, Tang C, Liu Y, Ma F, Zhang X, Zhang J (2018) Triterpenes and soluble polysaccharide changes in lingzhi or reishi medicinal mushroom, *Ganoderma lucidum* (Agaricomycetes), during fruiting growth. Int J Med Mushrooms 20(9):859–871

10. Zhou S, Ma F, Zhang X, Zhang J (2016) Carbohydrate changes during growth and fruiting in Pleurotus ostreatus. Fungal Biol 120(6–7):852–861
11. Yan X (2017) Short-chain polysaccharide analysis in ethanol-water solutions. J AOAC Int 100(4):1134–1136
12. Zhou S, Zhang J, Yang Y, Liu Y, Wu D, Zhang H, Tang Q (2012) Effect of particle size on the extraction of polysaccharide and triterpenes from finely ground *Ganoderma lucidum* fruit bodies. Acta Edulis Fungi 18(3):63–66
13. Zhu Y, Li Q, Mao G, Zou Y, Feng W, Zheng D, Wang W, Zhou L, Zhang T, Yang J, Yang L, Wu X (2014) Optimization of enzyme-assisted extraction and characterization of polysaccharides from Hericium erinaceus. Carbohydr Polym 101:606–613
14. Yin C, Fan X, Fan Z, Shi D, Gao H (2018) Optimization of enzymes-microwave-ultrasound assisted extraction of *Lentinus edodes* polysaccharides and determination of its antioxidant activity. Int J Biol Macromol 111:446–454
15. Huang S, Li J, Wang Z, Pan H, Chen J, Ning Z (2010) Optimization of alkaline extraction of polysaccharides from *Ganoderma lucidum* and their effect on immune function in mice. Molecules 15(5):3694
16. Chen TQ, Wu JG, Kan YJ, Yang C, Wu YB, Wu JZ (2018) Antioxidant and hepatoprotective activities of crude polysaccharide extracts from lingzhi or reishi medicinal mushroom, *Ganoderma lucidum* (Agaricomycetes), by ultrasonic-circulating extraction. Int J Med Mushrooms 20(6):581–593
17. Kang Q, Chen S, Li S, Wang B, Liu X, Hao L, Lu J (2019) Comparison on characterization and antioxidant activity of polysaccharides from *Ganoderma lucidum* by ultrasound and conventional extraction. Int J Biol Macromol 124:1137–1144
18. Smiderle FR, Morales D, Gil-Ramírez A, Jesus LID, Gilbert-López B, Iacomini M, Soler-Rivas C (2017) Evaluation of microwave-assisted and pressurized liquid extractions to obtain β- d -glucans from mushrooms. Carbohydr Polym 156:165–174
19. Chaplin M, Kennedy J (1994) Carbohydrate analysis, 2nd edn. Oxford University Press, Oxford
20. Zeng X, Li P, Chen X, Kang Y, Xie Y, Li X, Xie T, Zhang Y (2019) Effects of deproteinization methods on primary structure and antioxidant activity of *Ganoderma lucidum* polysaccharides. Int J Biol Macromol 126:867–876
21. Yuan H, Zhang J, Jia W, Yang Y, Ye L, Tang Q, Liu Y (2009) Study on decoloration of the low molecular weight *Ganoderma lucidum* polysaccharides by resin. Sci Technol Food Ind 30(3):204–206
22. Luo X, Tang Q, Zhang J, Zhou S, Wu D, Liu Y, Wang Q, Qiao Y, Zhao M (2011) Optimization of decolorization of *Ganoderma lucidum* polysaccharides by macroporous resin adsorption method. Food Sci 32(16):5–10
23. Yan JK, Wang YY, Qiu WY, Shao N (2017) Three-phase partitioning for efficient extraction and separation of polysaccharides from Corbicula fluminea. Carbohydr Polym 163:10–19
24. Liu Z, Yu D, Li L, Liu X, Zhang H, Sun W, Lin CC, Chen J, Chen Z, Wang W, Jia W (2019) Three-phase partitioning for the extraction and purification of polysaccharides from the immunomodulatory medicinal mushroom inonotus obliquus. Molecules 24(3):403
25. Habijanic J, Berovic M, Boh B, Plankl M, Wraber B (2015) Submerged cultivation of *Ganoderma lucidum* and the effects of its polysaccharides on the production of human cytokines TNF-alpha, IL-12, IFN-gamma, IL-2, IL-4, IL-10 and IL-17. New Biotechnol 32(1):85–95
26. Lei Y, Wang X, Liu Z, Wu X (2001) Purification of polysaccharide peptide of Coriolus versicolor in affinity chromatography. Chin Tradit Pat Med 23(1):8–10
27. Zhang J, Wang G, Li H, Zhuang C, Mizuno T, Ito H, Mayuzumi H, Okamoto H, Li J (1994) Antitumor active protein-containing glycans from the Chinese Mushroom Songshan Lingzhi, Ganoderma tsugae Mycelium. Biosci Biotechnol Biochem 58(7):1202–1205
28. Lin Y, Ma B, Wei H, Liu T, Lei L (2004) Purification technique of Bozhi glycopeptides by affinity chromatography. Biotechnology 14(z1):31–32

29. Huie CW, Di X (2004) Chromatographic and electrophoretic methods for Lingzhi pharmacologically active components. J Chromatogr B 812(1–2):241–257
30. Wasser S (2002) Medicinal mushrooms as a source of antitumor and immunomodulating polysaccharides. Appl Microbiol Biotechnol 60(3):258–274
31. Sassaki GL, Souza LM, Serrato RV, Cipriani TR, Gorin PAJ, Iacomini M (2008) Application of acetate derivatives for gas chromatography–mass spectrometry: novel approaches on carbohydrates, lipids and amino acids analysis. J Chromatogr 1208(1):215–222
32. Cui WS (2005) Food carbohydrates: chemistry, physical properties, and applications. CRC Press Inc, Boca Raton
33. Ruthes AC, Smiderle FR, Iacomini M (2015) d -Glucans from edible mushrooms: a review on the extraction, purification and chemical characterization approaches. Carbohydr Polym 117(117):753–761
34. Synytsya A, Novák M (2013) Structural diversity of fungal glucans. Carbohydr Polym 92(1):792–809
35. Bohn JA, BeMiller JN (1995) (1→3)-β-d-Glucans as biological response modifiers: a review of structure-functional activity relationships. Carbohydr Polym 28(1):3–14
36. Liu Y, Zhang J, Tang Q, Yang Y, Guo Q, Wang Q, Wu D, Cui SW (2014) Physicochemical characterization of a high molecular weight bioactive β-d-glucan from the fruiting bodies of *Ganoderma lucidum*. Carbohydr Polym 101(0):968–974
37. Liu Y, Tang Q, Zhang J, Xia Y, Yang Y, Wu D, Cui SW (2018) Triple helix conformation of β-D-glucan from *Ganoderma lucidum* and effect of molecular weight on its immunological activity. Int J Biol Macromol 114:1064–1070
38. Bao X, Duan J, Fang X, Fang J (2001) Chemical modifications of the (1→3)-α-d-glucan from spores of *Ganoderma lucidum* and investigation of their physicochemical properties and immunological activity. Carbohydr Res 336(2):127–140
39. Bao X, Fang J (2001) Isolation and structural determination of a glucan from the spores of *Ganoderma lucidum*. Acta Bot Sin 43(3):312–315
40. Bao X, Liu C, Fang J, Li X (2001) Structural and immunological studies of a major polysaccharide from spores of *Ganoderma lucidum* (Fr.) Karst. Carbohydr Res 332(1):67–74
41. Guo L, Xie J, Ruan Y, Zhou L, Zhu H, Yun X, Jiang Y, Lü L, Chen K, Min Z, Wen Y, Gu J (2009) Characterization and immunostimulatory activity of a polysaccharide from the spores of *Ganoderma lucidum*. Int Immunopharmacol 9(10):1175–1182
42. Li YQ, Fang L, Zhang KC (2007) Structure and bioactivities of a galactose rich extracellular polysaccharide from submergedly cultured *Ganoderma lucidum*. Carbohydr Polym 68(2):323–328
43. Ye L, Zhang J, Ye X, Tang Q, Liu Y, Gong C, Du X, Pan Y (2008) Structural elucidation of the polysaccharide moiety of a glycopeptide (GLPCW-II) from *Ganoderma lucidum* fruiting bodies. Carbohydr Res 343(4):746–752
44. Pan D, Wang L, Chen C, Teng B, Wang C, Xu Z, Hu B, Zhou P (2012) Structure characterization of a novel neutral polysaccharide isolated from *Ganoderma lucidum* fruiting bodies. Food Chem 135(3):1097–1103
45. Bao XF, Wang XS, Dong Q, Fang JN, Li XY (2002) Structural features of immunologically active polysaccharides from *Ganoderma lucidum*. Phytochemistry 59(2):175–181
46. Chang YW, Lu TJ (2004) Molecular characterization of polysaccharides in hot-water extracts of Ganoderma lucidum fruiting bodies. J Food Drug Anal 12(1):59–67
47. Wang J, Yuan Y, Yue T (2014) Immunostimulatory activities of β-d-glucan from *Ganoderma Lucidum*. Carbohydr Polym 102:47–54
48. Han XQ, Yue GL, Yue RQ, Dong CX, Chan CL, Ko CH, Cheung WS, Luo KW, Dai H, Wong CK (2014) Structure elucidation and immunomodulatory activity of a beta glucan from the fruiting bodies of *Ganoderma sinense*. PLoS One 9(7):e100380
49. Amaral AE, Carbonero ER, Simão RDCG, Kadowaki MK, Sassaki GL, Osaku CA, Gorin PAJ, Lacomini M (2008) An unusual water-soluble β-glucan from the basidiocarp of the fungus *Ganoderma resinaceum*. Carbohydr Polym 72(3):473–478

50. Li N, Yan C, Hua D, Zhang D (2013) Isolation, purification, and structural characterization of a novel polysaccharide from *Ganoderma capense*. Int J Biol Macromol 57(6):285–290
51. Ye L, Zheng X, Zhang J, Tang Q, Yang Y, Wang X, Li J, Liu Y, Pan Y (2011) Biochemical characterization of a proteoglycan complex from an edible mushroom *Ganoderma lucidum* fruiting bodies and its immunoregulatory activity. Food Res Int 44(1):367–372
52. Liu W, Wang HY, Pang XB, Yao WB, Gao XD (2010) Characterization and antioxidant activity of two low-molecular-weight polysaccharides purified from the fruiting bodies of *Ganoderma lucidum*. Int J Biol Macromol 46(4):451–457
53. Huang SQ, Li JW, Li YQ, Wang Z (2011) Purification and structural characterization of a new water-soluble neutral polysaccharide GLP-F1-1 from *Ganoderma lucidum*. Int J Biol Macromol 48(1):165–169
54. Zhang H, Nie S, Cui SW, Xu M, Ding H, Xie M (2017) Characterization of a bioactive polysaccharide from *Ganoderma atrum* : re-elucidation of the fine structure. Carbohydr Polym 158:58–67
55. Ye L, Zhang J, Zhou K, Yang Y, Zhou S, Jia W, Hao R, Pan Y (2008) Purification, NMR study and immunostimulating property of a fucogalactan from the fruiting bodies of *Ganoderma lucidum*. Planta Med 74(14):1730–1734
56. Zhang H, Nie S-P, Yin J-Y, Wang Y-X, Xie M-Y (2014) Structural characterization of a heterogalactan purified from fruiting bodies of *Ganoderma atrum*. Food Hydrocoll 36:339–347
57. Ye LB, Li JR, Zhang JS, Pan YJ (2010) NMR characterization for polysaccharide moiety of a glycopeptide. Fitoterapia 81(2):93–96
58. Ye LB, Zhang JS, Yan Y, Shuai Z, Liu YF, Tang QJ, Du XJ, Hui C, Pan YJ (2009) Structural characterisation of a heteropolysaccharide by NMR spectra. Food Chem 112(4):962–966
59. Zhang J, Tang Q, Zimmerman Kordmann M, Reutter W, Fan H (2002) Activation of B lymphocytes by GLIS, a bioactive proteoglycan from *Ganoderma lucidum*. Life Sci 71(6):623–638
60. Zhang J, Tang Q, Zhou C, Jia W, Silva LD, Long DN, Reutter W, Fan HJLS (2010) GLIS, a bioactive proteoglycan fraction from *Ganoderma lucidum*, displays anti-tumour activity by increasing both humoral and cellular immune response. Life Sci 87(19):628–637
61. Ji Z, Tang Q, Zhang J, Yang Y, Jia W, Pan Y (2007) Immunomodulation of RAW264.7 macrophages by GLIS, a proteopolysaccharide from *Ganoderma lucidum*. J Ethnopharmacol 112(3):445–450
62. Zhang K, Liu Y, Zhao X, Tang Q, Dernedde J, Zhang J, Fan H (2018) Anti-inflammatory properties of GLPss58, a sulfated polysaccharide from *Ganoderma lucidum*. Int J Biol Macromol 107:486–493
63. Shao B-M, Dai H, Xu W, Lin Z-B, Gao X-M (2004) Immune receptors for polysaccharides from *Ganoderma lucidum*. Biochembiophysrescommun 323(1):133–141
64. Chien CM, Cheng J-L, Chang W-T, Tien M-H, Tsao C-M, Chang Y-H, Chang H-Y, Hsieh J-F, Wong C-H, Chen S-T (2004) Polysaccharides of *Ganoderma lucidum* alter cell immunophenotypic expression and enhance CD56+ NK-cell cytotoxicity in cord blood. Bioorg Med Chem 12(21):5603–5609
65. Wang H, Yu Q, Nie SP, Xiang QD, Zhao MM, Liu SY, Xie MY, Wang SQ (2017) Polysaccharide purified from *Ganoderma atrum* induced activation and maturation of murine myeloid-derived dendritic cells. J Food Chem Toxicol 108:478–485
66. CAo L-Z, Lin Z-B (2002) Regulation on maturation and function of dendritic cells by *Ganoderma lucidum* polysaccharides. Immunol Lett 83(3):163–169
67. Zhang J, Gao X, Pan Y, Xu N, Jia L (2016) Toxicology and immunology of *Ganoderma lucidum* polysaccharides in Kunming mice and Wistar rats. Int J Biol Macromol 85:302–310
68. Pi CC, Chu CL, Lu CY, Zhuang YJ, Wang CL, Yu YH, Wang HY, Lin CC, Chen CJ (2014) Polysaccharides from *Ganoderma formosanum* function as a Th1 adjuvant and stimulate cytotoxic T cell response in vivo. Vaccine 32(3):401–408
69. Jin M, Zhu Y, Shao D, Zhao K, Xu C, Li Q, Yang H, Huang Q, Shi J (2017) Effects of polysaccharide from mycelia of *Ganoderma lucidum* on intestinal barrier functions of rats. Int J Biol Macromol 94(Pt A):1–9

70. Kubota A, Kobayashi M, Sarashina S, Takeno R, Okamoto K, Narumi K, Furugen A, Suzuki Y, Takahashi N, Iseki K (2018) Reishi mushroom *Ganoderma lucidum* modulates IgA production and alpha-defensin expression in the rat small intestine. J Ethnopharmacol 214:240–243
71. Huang SQ, Ning ZX (2010) Extraction of polysaccharide from Ganoderma lucidum and its immune enhancement activity. Int J Biol Macromol 47(3):336–341
72. Ferreira ICFR, Heleno SA, Reis FS, Stojkovic D, Queiroz MJRP, Vasconcelos MH, Sokovic M (2015) Chemical features of Ganoderma polysaccharides with antioxidant, antitumor and antimicrobial activities. Phytochemistry 114:38–55
73. Zhao L, Dong Y, Chen G, Hu Q (2010) Extraction, purification, characterization and antitumor activity of polysaccharides from *Ganoderma lucidum*. Carbohydr Polym 80(3):783–789
74. Zhang S, Pang G, Chen C, Qin J, Yu H, Liu Y, Zhang X, Song Z, Zhao J, Wang FJCP (2019) Effective cancer immunotherapy by *Ganoderma Lucidum* polysaccharide-gold nanocomposites through dendritic cell activation and memory T cell response. Carbohydr Polym 205:192–202
75. Gao Y, Tang W, Dai X, Gao H, Chen G, Ye J, Chan E, Koh HL, Li X, Zhou S (2005) Effects of water-soluble *Ganoderma lucidum* polysaccharides on the immune functions of patients with advanced lung cancer. J Med Food 8(2):159–168
76. Joseph S, Sabulal B, George V, Antony KR, Janardhanan KK (2011) Antitumor and anti-inflammatory activities of polysaccharides isolated from *Ganoderma lucidum*. Acta Pharma 61(3):335–342
77. Zhang S, Nie S, Huang D, Li W, Xie M (2013) Immunomodulatory effect of *Ganoderma atrum* polysaccharide on CT26 tumor-bearing mice. Food Chem 136(3–4):1213–1219
78. Sun X, Zhao C, Pan W, Wang J, Wang W (2015) Carboxylate groups play a major role in antitumor activity of *Ganoderma applanatum* polysaccharide. Carbohydr Polym 123:283–287
79. Zhang S, Nie S, Huang D, Huang J, Feng Y, Xie M (2014) *Ganoderma atrum* polysaccharide evokes antitumor activity via cAMP-PKA mediated apoptotic pathway and down-regulation of Ca2+/PKC signal pathway. Food Chem Toxicol 68:239–246
80. Liu Y, Li Y, Zhang W, Sun M, Zhang Z (2019) Hypoglycemic effect of inulin combined with *ganoderma lucidum* polysaccharides in T2DM rats. J Funct Foods 55:381–390
81. Yang Z, Chen C, Zhao J, Xu W, He Y, Yang H, Zhou P (2017) Hypoglycemic mechanism of a novel proteoglycan, extracted from *Ganoderma lucidum*, in hepatocytes. Eur J Pharmacol 820:77–85
82. Ma H-T, Hsieh J-F, Chen S-T (2015) Anti-diabetic effects of *Ganoderma lucidum*. Phytochemistry 114:109–113
83. Xiao C, Wu Q, Zhang J, Xie Y, Cai W, Tan J (2016) Antidiabetic activity of *Ganoderma lucidum* polysaccharides F31 down-regulated hepatic glucose regulatory enzymes in diabetic mice. J Ethnopharmacol 196:47–57
84. Gao Y, Lan J, Dai X, Ye J, Zhou S (2004) A phase I/II study of Ling Zhi mushroom *Ganoderma lucidum* (W.Curt.:Fr.) Lloyd (Aphyllophoromycetideae) extract in patients with coronary heart disease. Int J Med Mushrooms 6(4):327–334
85. Zhang HN, He JH, Yuan L, Lin ZB (2003) In vitro and in vivo protective effect of *Ganoderma lucidum* polysaccharides on alloxan-induced pancreatic islets damage. Life Sci 73(18):2307–2319
86. Liu YJ, Du JL, Cao LP, Jia R, Shen YJ, Zhao CY, Xu P, Yin GJ (2015) Anti-inflammatory and hepatoprotective effects of *Ganoderma lucidum* polysaccharides on carbon tetrachloride-induced hepatocyte damage in common carp (Cyprinus carpio L.). Int Immunopharmacol 25(1):112–114
87. Jin H, Xu J, Tao T-T, Huang H-J, Jin, Feng J (2013) Protective effects of *Ganoderma lucidum* spore on cadmium hepatotoxicity; in mice. Food Chem Toxicol 52(2):171–175
88. Fan S, Huang X, Wang S, Li C, Zhang Z, Xie M, Nie S (2018) Combinatorial usage of fungal polysaccharides from Cordyceps sinensis and *Ganoderma atrum* ameliorate drug-induced liver injury in mice. Food Chem Toxicol 119:66–72

89. Zhang G-L, Wang Y-H, Ni W, Teng H-L, Lin Z-B (2002) Hepatoprotective role of *Ganoderma lucidum* polysaccharide against BCG-induced immune liver injury in mice. World J Gastroenterol 8(4)):728–733
90. Wu X, Zeng J, Hu J, Liao Q, Zhou R, Zhang P, Chen Z (2013) Hepatoprotective effects of aqueous extract from Lingzhi or Reishi medicinal mushroom *Ganoderma lucidum* (higher basidiomycetes) on α-amanitin-induced liver injury in mice. Int J Med Mushrooms 15(4):383–391
91. Zhu K, Nie S, Li C, Lin S, Xing M, Li W, Gong D, Xie M (2013) A newly identified polysaccharide from *Ganoderma atrum* attenuates hyperglycemia and hyperlipidemia. Int J Biol Macromol 57:142–150
92. Chen WQ, Luo SH, Li HZ, Yang H (2005) Effects of *Ganoderma lucidum* polysaccharides on serum lipids and lipoperoxidation in experimental hyperlipidemic rats. Zhongguo Zhong Yao Za Zhi 30(17):1358–1360
93. Qian Y, Siwang W, Yanhua X, Jiyuan S, Jianbo W (2010) HPLC analysis of *Ganoderma lucidum* polysaccharides and its effect on antioxidant enzymes activity and Bax, Bcl-2 expression. Int J Biol Macromol 46(2):167–172
94. Jin M, Zhang H, Wang J, Shao D, Yang H, Huang Q, Shi J, Xu C, Zhao K (2019) Response of intestinal metabolome to polysaccharides from mycelia of *Ganoderma lucidum*. Int J Biol Macromol 122:723–731
95. Zhao C, Fan J, Liu Y, Guo W, Cao H, Xiao J, Wang Y, Liu B (2019) Hepatoprotective activity of *Ganoderma lucidum* triterpenoids in alcohol-induced liver injury in mice, an iTRAQ-based proteomic analysis. Food Chem 271:148–156
96. Gokce EC, Kahveci R, Atanur OM, Gãrer B, Aksoy N, Gokce A, Sargon MF, Cemil B, Erdogan B, Kahveci O (2015) Neuroprotective effects of *Ganoderma lucidum* polysaccharides against traumatic spinal cord injury in rats. Inj Int J Care Inj 46(11):2146–2155
97. Li W-J, Zhang X-Y, Wu R-T, Song Y-H, Xie M-Y (2018) *Ganoderma atrum* polysaccharide improves doxorubicin-induced cardiotoxicity in mice by regulation of apoptotic pathway in mitochondria. Carbohydr Polym 202:581–590
98. Chen J, Shi Y, He L, Hao H, Wang B, Zheng Y, Hu C (2016) Protective roles of polysaccharides from *Ganoderma lucidum* on bleomycin-induced pulmonary fibrosis in rats. Int J Biol Macromol 92:278–281
99. Zhang H, Cui SW, Nie SP, Chen Y, Wang YX, Xie MY, Fibre D (2016) Identification of pivotal components on the antioxidant activity of polysaccharide extract from *Ganoderma atrum*. J Bioact Carbohydr 7(2):9–18

Chapter 5
Quantitative Analysis of Components in *Ganoderma*

Ruoyun Chen and Jie Kang

Abstract The components in *Ganoderma*, including triterpenes, polysaccharides, nucleosides or nucleobases, steroids, and ergothioneine, were quantitatively analyzed. Among them, the quantitative analysis of triterpenes and polysaccharides were recorded in US Pharmacopeia. HPLC-UV and HPLC-MS are the most popular quantitative methods for all of these compounds.

Keywords *Ganoderma* · Triterpenes · Polysaccharides · Nucleosides or nucleobases · Steroids · Ergothioneine · Quantitative analysis

5.1 Introduction

The analytical methods (HPLC-UV) for quantitative analysis of triterpenes and polysaccharides in *Ganoderma* were recorded in US Pharmacopeia [1]. The other methods for quantitative analysis of triterpenes (HPLC-MS) [2], polysaccharides (phenol-sulfuric acid, mid-infrared, and near-infrared spectroscopy) [3–5], sterols (HPLC-UV) [6, 7], nucleosides or nucleobases (HPLC-UV, HPLC-MS and CE-MS) [8–10], and ergothioneine (HPLC-UV) [11] in *Ganoderma* were summarized, too. The contents of these components in *Ganoderma* from different species, parts, and origins were varied.

R. Chen · J. Kang (✉)
Department of Natural Product Chemistry, Institute of Materia Medica, Chinese Academy of Medical Sciences & Peking Union Medical College, Beijing, China
e-mail: jiekang@imm.ac.cn

Z. Lin, B. Yang (eds.), *Ganoderma and Health*, Advances in Experimental Medicine and Biology 1181, https://doi.org/10.1007/978-981-13-9867-4_5

5.2 The Analytical Methods for Quantitative Analysis of Triterpenes in *Ganoderma*

5.2.1 The Method of Quantitative Analysis of Triterpenes in Ganoderma *Recorded in US Pharmacopeia [1]*

The mobile phase consisted of solvent A (0.075% phosphoric acid in water) and solvent B (acetonitrile). The gradient mobile phase was as follows: 0~3 min, 20~26.5% B; 3~34 min, 26.5~26.5% B; 34~52 min, 26.5~38.5% B; 52~53 min, 38.5~20% B; and 53~58 min, 20~20%. The column was 2.1 mm × 15 cm (1.8 μm). The flow rate was 0.4 mL/min. The temperature of column compartment was kept at 25 °C, and the peaks were detected at a wavelength of 257 nm.

Standard Solution A 0.1 mg/mL of USP ganoderic acid A in methanol. Sonicate to dissolve if necessary.

Standard Solution B Sonicate 40 mg of USP *G. lucidum* fruiting body powdered extract in 5 mL of alcohol, and centrifuge. Pass through a nylon filter of 0.2 μm pore size, and discard the initial 1 mL of the filtrate.

Sample Solution Transfer 2.0 g of *G. lucidum* fruiting body, finely powdered and accurately weighed, to a 200 mL round-bottom flask, add 75 mL of alcohol, attach a condenser, reflux for 45 min, cool, and filter. Rinse the flask with two 10 mL portions of alcohol, and filter, combining the washes and the filtrate. Evaporate to dryness under reduced pressure, and dissolve the residue in about 20 mL of alcohol. Transfer the solution to a 25 mL volumetric flask, dilute with alcohol to volume, and mix well. Pass through a nylon filter of 0.2 μm pore size, and discard the initial 1 mL of the filtrate.

Analysis Using the chromatograms of standard solution A, standard solution B, and the reference chromatogram provided with the lot of USP *G. lucidum* fruiting body powdered extract being used, identify all specified ganoderic and ganoderenic acids in the *Sample Solution* chromatogram. The approximate relative retention times, with respect to ganoderic acid A, are provided in Table 5.1.

Separately calculate the percentages of each triterpenoic acid in the portion of *G. lucidum* fruiting body taken:

$$\text{Result} = (r_u / r_s) \times Cs \times (V / W) \times F \times 100$$

r_u = peak area of the relevant analyte in the sample solution,
r_s = peak area of ganoderic acid in the standard solution A,
Cs = concentration of USP ganoderic acid A in standard solution A
V = volume of the sample solution (mL)

Table 5.1 The approximate relative retention times and factors of ten ganoderic or ganoderenic acids, with respect to ganoderic acid A

Analyte	Relative retention time (min)	Relative response factor
Ganoderenic acid C	0.36	0.51
Ganoderic acid C_2	0.42	1.05
Ganoderic acid G	0.56	1.18
Ganoderenic acid B	0.60	0.45
Ganoderic acid B	0.66	1.10
Ganoderic acid A	1.00	1.00
Ganoderic acid H	1.05	1.54
Ganoderenic acid D	1.25	0.51
Ganoderic acid D	1.33	1.08
Ganoderic acid F	1.54	1.45

W = weight of *G. lucidum* fruiting body taken to prepare the sample solution (mg)
F = relative response factor, with *respect* to ganoderic acid A (see Table 5.1)

Calculate the sum of the percentages of all specified triterpenoic acids.

5.2.2 *The Other Methods of Quantitative Analysis of Triterpenes in* Ganoderma

5.2.2.1 The Standards of Triterpenes in *Ganoderma*

A total of 53 triterpenes were quantitatively analyzed in *G. lucidum*, *G. lingzhi* and *G. sinense*.

Thirty-seven triterpenes, which were ganoderic acids A (**1**), AM1 (**2**), B (**3**), C2 (**4**), D (**5**), DM (**6**), E (**7**), F (**8**), G (**9**), H (**10**), K (**11**), TR (**12**), α (**13**), η (**14**), ε (**15**), and θ (**16**); ganoderiols A (**17**), F (**18**); ganoderol A (**19**); ganoderal A (**20**); ganoderate G (**21**); methyl ganoderate D (**22**); ganodermadiol (**23**); ganodermatriol (**24**); ganodermanontriol (**25**); ganodermanondiol (**26**); lucidumols A (**27**) and B (**28**); ganolucidic acid A (**29**); lucidenic acids A (**30**), E_2 (**31**), and N (**32**); methyl lucidenate A (**33**) and E_2 (**34**); butyl lucidenate E_2 (**35**); 20-OH lucidenic acid A (**36**); and 20-OH lucidenic acid N (**37**), were quantitatively determined in *G. lucidum* [1, 2, 6, 7, 12–17], respectively.

Seventeen triterpenes, ganoderic acids A (**1**), B (**3**), C2 (**4**), D (**5**), DM (**6**), F (**8**), G (**9**), and T (**38**); ganoderenic acids A (**39**), B (**40**), D (**41**), and F (**42**); ganoderiols B (**43**) and F (**18**); ganoderal A (**44**); ganodermanontriol (**25**), and lucidenic acid A (**30**), were quantitatively determined in *G. lingzhi* [13], respectively.

Six triterpenes, ganoderic acid A (**1**); ganoderiols A (**17**), D (**45**), and F (**18**), ganodermanontriol (**25**), lucidumol A (**27**), were quantitatively determined in *G. sinense* [2, 13, 15], respectively.

5.2.2.2 The Extraction Methods of Triterpenes in *Ganoderma*

The powdered fruiting bodies or spores were extracted by heating, ultrasonication, or pressurized liquid extraction with chloroform, ethyl acetate, methanol, or ethyl alcohol [1, 2, 6, 7, 12–17].

5.2.2.3 The Analytical Methods of Triterpenes in *Ganoderma*

HPLC analyses were performed on Waters 600, Waters 2695, Tosoh CCP 8020, Shimadzu LC 20A, Agilent 1100 HPLC, or Agilent 1290 UPLC system, and the UV detection wavelength was set at 256, 254, 252, or 243 nm, respectively [1, 2, 6, 7, 12–17].

For HPLC/MS analyses, ESI/MS was chosen as the ion source. Quantification was achieved in SRM mode with a combination of a specific precursor ion and a unique product ion for most target triterpenes [2].

A Zorbax RRHD Eclipse Plus C_{18} (2.1 × 50 mm, 1.8 μm), Zorbax SB C_{18} (4.6 × 250 mm, 5 μm), Fortis Speed Core-C-18 (4.6 × 150 mm, 2.6 μm), Alltima C_{18} (4.6 × 150 mm, 5 μm), or TSK gel ODS-80 Ts (4.6 × 150 mm, 5 μm) column was used in analyses, respectively [1, 2, 6, 7, 12–17].

The mobile phase was composed of lipophilic and hydrophilic solvents. Methanol or acetonitrile were used as lipophilic solvents. Water, 0.01% acetic acid-water, 1% acetic acid-water, 2% acetic acid-water, 0.1% formic acid-water, 0.04% formic acid-water, and 0.03% phosphoric acid-water were used as hydrophilic solvents, respectively [1, 2, 6, 7, 12–17].

5.2.2.4 The Total Contents of Triterpenes in *Ganoderma*

Mixed standards were used in HPLC-UV or HPLC-MS analysis, and internal standards of cholic acid and hydrocortisone were added for HPLC-MS analyses of triterpene acids and triterpene alcohols [2], respectively.

A paper reported one triterpene (ganoderic acid B) used to quantitatively determine total triterpene acids in *G. lucidum*, *G. sinense* and *G. tsugae* [16].

The total contents of triterpenes were varied and summarized in Table 5.2. In general, more contents of triterpenes were found in *G. lucidum* and *G. lingzhi* than that of *G. sinense*. The other species of *Ganoderma*, including *G. amboinense*, *G. sessile*, *G. atrum*, *G. tropicum*, *G. resinaceum*, *G. applanatum*, and *G. crebrossinaceum*, were less analyzed [1, 2, 6, 7, 12–17].

Table 5.2 The contents and types of total triterpenes in different batches of *Ganoderma* (μg/g)

No.	*Ganoderma*	Origin	Total triterpenes (μg/g)	Types of triterpenes	References	No.	*Ganoderma*	Origin	Total triterpenes (μg/g)	Types of triterpenes	References
1	*G. lucidum*-F[a]-1	–	3875.5	17	[12]	53	*G. lucidum*-F-53	Bac Giang, Vietnam	6178.5	8	[7]
2	*G. lucidum*-F-2	–	2030.2	16	[12]	54	*G. lucidum*-F-54	Vietnam Academy of Agricultural Sciences	6250.9	8	[7]
3	*G. lucidum*-F-3	–	2221.2	17	[12]	55	*G. lucidum*-F-55	–	1876.0	7	[7]
4	*G. lucidum*-F-4	–	2265.6	15	[12]	56	*G. lucidum*-F-56	–	1466.6	8	[7]
5	*G. lucidum*-F-5	–	3452.6	18	[12]	57	*G. lucidum*-F-57	–	1553.2	9	[7]
6	*G. lucidum*-F-6	–	1143.1	16	[12]	58	*G. lucidum*-F-58	Vietnam Academy of Agricultural Sciences	1525.5	8	[7]
7	*G. lucidum*-F-7	–	4007.9	16	[12]	59	*G. lingzhi*-F-59	Jilin, China	4807.8	16	[13]
8	*G. lucidum*-F-8	–	6276.0	18	[12]	60	*G. lingzhi*-F60	Shandong, China	3884.2	16	[13]
9	*G. lucidum*-F-9	–	5009.8	18	[12]	61	*G. lingzhi*-F-61	Shanghai, China	2652.1	16	[13]

(continued)

Table 5.2 (continued)

No.	*Ganoderma*	Origin	Total triterpenes (μg/g)	Types of triterpenes	References	No.	*Ganoderma*	Origin	Total triterpenes (μg/g)	Types of triterpenes	References
10	*G. lucidum*-spore-10	–	4046.8	18	[12]	62	*G. lingzhi*-F62	Shanghai, China	5088.9	16	[13]
11	*G. lucidum*-F-11	–	935.8	12	[12]	63	*G. lingzhi*-F-63	Shanghai, China	3906.8	15	[13]
12	*G. lucidum*-F-12	–	890.5	12	[12]	64	*G. lingzhi*-F-64	Shanghai, China	3389.4	16	[13]
13	*G. lucidum*-F-13	–	590.4	12	[12]	65	*G. lingzhi*-F-65	Shanghai, China	5352.8	16	[13]
14	*G. lucidum*-F-14	–	1759.2	14	[12]	66	*G. lingzhi*-F-66	Shanghai, China	3917.0	17	[13]
15	*G. lucidum*-F-15	–	846.6	13	[12]	67	*G. lingzhi*-F-67	Shanghai, China	5416.9	16	[13]
16	*G. lucidum*-F-16	–	813.0	13	[12]	68	*G. lingzhi*-F-68	Shanghai, China	2914.3	16	[13]
17	*G. lucidum*-F-17	–	1268.4	14	[12]	69	*G. lingzhi*-F-69	Shanghai, China	4008.9	17	[13]
18	*G. lucidum*-F-18	–	1202.8	14	[12]	70	*G. lingzhi*-F-70	Shanghai, China	3634.9	17	[13]
19	*G. lucidum*-F-19	–	3036.0	6	[15]	71	*G. lingzhi*-F-71	Shanghai, China	3880.4	16	[13]
20	*G. lucidum*-F-20	Fujian, China	6870.3	6	[15]	72	*G. lingzhi*-F-72	Shanghai, China	3589.6	16	[13]
21	*G. lucidum*-F-21	Fujian, China	11551.9	6	[15]	73	*G. lingzhi*-F-73	Shanghai, China	3787.6	17	[13]

No.	*Ganoderma*	Origin	Total triterpenes (μg/g)	Types of triterpenes	References	No.	*Ganoderma*	Origin	Total triterpenes (μg/g)	Types of triterpenes	References
22	*G. lucidum*-F-22	Fujian, China	6526.4	6	[15]	74	*G. lingzhi*-F-74	Shanghai, China	3303.1	16	[13]
23	G. lucidum-F-23	Fujian, China	4501.4	6	[15]	75	*G. lingzhi*-F-75	Shanghai, China	4426.6	17	[13]
24	G. lucidum-F-24	Fujian, China	4492.2	6	[15]	76	*G. lingzhi*-F-76	Shanghai, China	3210.7	16	[13]
25	*G. lucidum*-F-25	Fujian, China	2803.1	6	[15]	77	*G. sinense*-F-3	–	341.8	15	[13]
26	*G. lucidum*-F-26	Fujian, China	5473.0	6	[15]	78	*G. sinense*-F-4	–	294.7	8	[13]
27	*G. lucidum*-F-27	Fujian, China	5686.7	6	[15]	79	*G. sinense*-F-5	Yunnan, China	0	5	[6]
28	*G. lucidum*-F-28	Fujian, China	5235.8	6	[15]	80	*G. sinense*-F-6	–	0	5	[6]
29	*G. lucidum*-F-29	Fujian, China	8140.9	6	[15]	81	*G. sinense*-F-7	–	0	5	[6]
30	*G. lucidum*-F-30	Anhui, China	8181.4	6	[15]	82	*G. sinense*-F-8	Sichuan, China	0	5	[6]
31	*G. lucidum*-F-31	Xinjiang, China	5229.2	6	[15]	83	*G. sinense*-F-9	–	0	5	[6]
32	*G. lucidum*-F-32	Jiangsu, China	4373.7	6	[15]	84	*G. sinense*-F-10	Guizhou, China	520.2	4	[15]
33	*G. lucidum*-F-33	Yunnan, China	6909.6	6	[15]	85	*G. sinense*-F-11	Xinjiang, China	73.1	2	[15]

(continued)

Table 5.2 (continued)

No.	*Ganoderma*	Origin	Total triterpenes (μg/g)	Types of triterpenes	References	No.	*Ganoderma*	Origin	Total triterpenes (μg/g)	Types of triterpenes	References
34	*G. lucidum*-F-34	Hebei, China	3252.8	6	[15]	86	*G. sinense*-F-12	Guizhou, China	152.5	3	[15]
35	*G. lucidum*-F-35	Liaoning, China	3338.5	6	[15]	87	*G. sinense*-F-13	Guangdong, China	117.7	4	[15]
36	*G. lucidum*-F-36	Tibet, China	2108.9	6	[15]	88	*G. sinense*-F-14	Fujian, China	687.8	6	[15]
37	*G. lucidum*-F-37	Jilin, China	4764.8	6	[15]	89	*G. sinense*-F-15	Guangxi, China	1918.1	5	[15]
38	*G. lucidum*-F-38	Beijing, China	5195.9	6	[15]	90	*G. sinense*-F-16	Zhejiang, China	37.8	1	[15]
39	*G. lucidum*-F-39	Guangxi, China	6810.8	6	[15]	91	*G. amboinense*-F-1	Fujian, China	10513.8		[15]
40	*G. lucidum*-F-40	Hubei, China	1858.5	6	[15]	92	*G. amboinense*-F-2	Tibet, China	737.4		[15]
41	*G. lucidum*-F-41	Yunnan, China	1250	2	[6]	93	*G. amboinense*-F-3	Xinjiang, China	6526.3		[15]
42	*G. lucidum*-F-42	–	2010	5	[6]	94	*G. amboinense*-F-4	–	3264.4		[13]
43	*G. lucidum*-F-43	Sichuan, China	890	2	[6]	95	*G. sessile*-F-1	Sichuan, China	1380.3		[15]
44	*G. lucidum*-F-44	–	1530	2	[6]	96	*G. sessile*-F-2	Tibet, China	167.3		[15]
45	*G. lucidum*-F-45	Anhui, China	3880	6	[6]	97	*G. atrum*-F-1	Fujian, China	507.8		[15]
46	*G. lucidum*-F-46	Guangdong, China	570	1	[6]	98	*G. atrum*-F-2	–	33.5		[13]

No.	*Ganoderma*	Origin	Total triterpenes (μg/g)	Types of triterpenes	References	No.	*Ganoderma*	Origin	Total triterpenes (μg/g)	Types of triterpenes	References
47	*G. lucidum*-F-47	Jilin, China	22200.6	8	[2]	99	*G. tropicum*-F-1	Fujian, China	8974.1		[15]
48	*G. lucidum*-F-48	Anhui, China	2790.6	8	[2]	100	*G. resinaceum*-F-1	–	759.9		[13]
49	*G. lucidum*-F-49	Guangnam, Vietnam	3548.0	9	[7]	101	*G. applanatum*-F-1	–	379.9		[13]
50	*G. lucidum*-F-50	Guangnam, Vietnam	5522.6	8	[7]	102	*G. crebrossinaceum* -F-1	–	1849.7		[13]
51	*G. lucidum*-F-51	Guangnam, Vietnam	3319.9	8	[7]						
52	*G. lucidum*-F-52	Bac Giang, Vietnam	3427.9	8	[7]						

[a]F stands for fruiting body

5.3 The Analytical Methods for Quantitative Analysis of Sterols in *Ganoderma*

Up to now, only one sterol, ergosterol, was quantitatively analyzed in *Ganoderma*.

Analytical HPLC was carried out on an Agilent Series 1100 or a LC 20A system. A Zorbax XDB C18 column (4.6 × 250 mm, 5 μm, Agilent Technologies, Inc., Santa Clara, CA, USA) was used. Two gradient solution systems consisted of water (A) and methanol (B) or 0.1% phosphoric acid in water (A) and acetonitrile (B) were used in analysis. The flow rate was 1.0 or 0.5 mL/min, the system was operated at 25 or 40 °C, and the detection wavelengths were set at 254 or 243 nm [6, 7].

The contents of ergosterol in *Ganoderma* were shown in Table 5.3 [6, 7].

5.4 The Analytical Methods for Quantitative Analysis of Polysaccharides in *Ganoderma*

5.4.1 The Method of Quantitative Analysis of Polysaccharides in Ganoderma *Recorded in US Pharmacopeia [1]*

The mobile phase consisted of solvent A (0.05 M phosphate buffer, pH 6.0) and solvent B (acetonitrile). The gradient mobile phase was as follows: 0~30 min, 16~17.5% B; 30~55 min, 17.5~19.0% B; 55~60 min, 19.0~19.0% B; and 60~61 min, 19.0~16.0% B. The column was 4.6 mm × 25 cm (5 μm). The flow rate was 1.0 mL/min. The temperature of column compartment was kept at 35 °C, and the peaks were detected at a wavelength of 250 nm.

Reagent 0.1 M solution of 1-phenyl-3-methyl-5-pyrazolone in methanol.

Internal Standard Solution 0.5 mg/mL of D-lyxose in water.

Standard Stock Solution Composite solution containing 0.20 mg/mL each of USP mannose, USP D-glucuronic acid, and USP galactose, 2.0 mg/mL of USP dextrose, and 0.10 mg/mL of USP L-fucose in water.

Standard Solution Combine 0.125 mL of the standard stock solution with 0.125 mL of the internal standard solution, 0.300 mL of 0.15 M sodium hydroxide solution, and 0.50 mL of regent in a capped reaction vial. Seal the vial, heat at 70 °C for 30 min, and cool to room temperature. Add to the vial 0.300 mL of 0.15 M hydrochloric acid and 0.65 mL of water, mix well, and pass through a nylon filter of 0.45 μM or finer pore size.

Sample Solution Transfer 2.0 g of *G. lucidum* fruiting body, finely powdered and accurately weighed, to a 200 mL round-bottom flask, add 60 mL of alcohol, and

Table 5.3 The contents of ergosterol in batches of fruiting bodies of *Ganoderma* (*μ*g/g)

No.	**1**	**2**	**3**	**4**	**5**	**6**	**7**	**8**	**9**	**10**
Ganoderma	*G. lucidum*	*G. lucidum*	*G. lucidum*	*G. lucidum*	*G. lucidum*	*G. lucidum*	*G. sinense*	*G. sinense*	*G. sinense*	*G. sinense*
Origin	Yunnan, China	–	Sichuan, China	–	Anhui, China	Guangdong, China	Yunnan, China	–	–	Sichuan, China
Ergosterol (μg/g)	1350	2300	1870	1840	2250	1720	880	1680	1660	2900
No.	**11**	**12**	**13**	**14**	**15**	**16**	**17**	**18**	**19**	**20**
Ganoderma	*G. sinense*	*G. lucidum*	*G. lucidum*	*G. lucidum*	*G. lucidum*	*G. applanatum*	*G. colossum*	*G. subresinosum*	*G.* sp.	*G. australe*
Origin	–	Guangnam, Vietnam	Guangnam, Vietnam	Guangnam, Vietnam	Guangnam, Vietnam	–	–	–	–	–
Ergosterol (μg/g)	930	135.1	194.6	795.9	647.8	112.7	221.5	158.3	116.3	148.3

allow to stand for 1 h. Attach a condenser, heat under reflux for 4 h, and filter immediately. Transfer the residue and the filter to the same 200 mL round-bottom flask, add 60 mL of water, heat under reflux for 3 h, and filter immediately. Rinse the flask with three 5 mL portions of water, and filter. Combine the filtrates and the rinsates in a 250 mL beaker, and evaporate on the water bath to dryness. Dissolve the residue in 5 mL of water, add 75 mL of alcohol, mix well, allow to stand at 4 °C for 12 h, and centrifuge at 4000 rpm for 30 min. Discard the supernatant, and dry the precipitate on a water bath. Dissolve the residue in hot water, quantitatively transfer to a 10 mL volumetric flask, cool to room temperature, and dilute 4000 rpm for 10 min. Accurately transfer 0.250 mL of the supernatant to a reaction vial, and add about 0.25 mL of 4 M trifluoroacetic acid. Seal the vial, heat at 110 °C for 4 h, cool to room temperature, add 0.5 mL of methanol, and evaporate to dryness at 60 °C under vacuum. Repeat the addition of 0.5 mL of methanol and subsequent evaporation three times. Add to the residue 0.125 mL of water, 0.125 mL of the internal standard solution, 0.300 mL of 0.15 M sodium hydroxide solution, and 0.50 mL of reagent. Seal the vial, heat at 70 °C for 30 min, and cool to room temperature. Hydrochloric acid (0.300 mL, 0.15 M) and 0.65 mL of water were added to the vial, mixed well, and passed through a nylon filter of 0.45 μm or finer pore size.

Analysis Using the chromatograms of standard solution and the reference chromatogram provided with the lot of USP *G. lucidum* fruiting body powdered extract being used, identify the individual derivatized monosaccharides at about the following relative retention times, with respect to dextrose: 0.48 for mannose, 0.58 for lyxose, 0.82 for D-glucuronic acid, 1.09 for galactose, and 1.35 for L-fucose.

Separately calculate the percentages of derivatized monosaccharides in the portion of *G. lucidum* fruiting body taken:

$$\text{Result} = \left(R_U / R_S\right) \times As \times \left(F / W\right) \times 100$$

R_U = peak response ratio of the relevant analyte to the internal standard in the sample solution
R_S = peak response ratio of the relevant analyte to the internal standard in the standard solution
As = amount of the relevant analyte in the aliquot of the standard solution subjected to derivatization (mg)
F = dilution factor to account for the sample aliquot submitted to derivatization (0.250 mL) relative to the volume of the sample solution (10.0 mL)
W = weight of *G. lucidum* fruiting body taken to prepare the sample solution (mg)

Calculate the sum of the percentages of mannose, D-glucuronic acid, dextrose, galactose, and L-fucose.

5.4.2 *The Other Method of Quantitative Analysis of Polysaccharides in* Ganoderma

The content of polysaccharides in *Ganoderma* was quantitatively determined by phenol-sulfuric acid method [1] or mid-infrared and near-infrared spectroscopy.

5.4.2.1 Phenol-Sulfuric Acid Method

1. **The Extraction Methods of Polysaccharides in *Ganoderma***

   Polysaccharides were isolated by the method proposed by Pillai with some modifications [3]. Ten grams of fungal material was extracted three times with 200 mL of hot water (85 °C) for 5 h. Water extracts were filtered, combined, and concentrated to 100 mL. Then 300 mL of chilled ethanol was added and it was left in a cold place (4 °C) for 24 h. The precipitate was collected after centrifugation (5000 rpm, 10 min), washed with ethanol, and dried. Crude polysaccharides were obtained.

2. **Determination of Total Sugar in Crude Polysaccharide**

   The total sugar in a sample of crude polysaccharide of *G. lucidum* was determined using a phenol-sulfuric acid method [3]. One milligram of each sample was dissolved in 10 mL of distilled water. One milliliter of this solution was mixed with 1 mL of 5% phenol solution and 5 mL of concentrated sulfuric acid. The mixture was shaken for 30 min and then the absorbance was measured at 490 nm. The total sugar was calculated based on the standard curve of glucose. All experiments were repeated three times. The standard deviation (SD) and relative standard deviation (RSD) value (the percentage of relative standard deviation) were calculated.

5.4.2.2 Mid-Infrared and Near-Infrared Spectroscopy [4, 5]

1. **Extraction and Purification of Polysaccharides in *Ganoderma***

   Fermented *Ganoderma* mycelia dried powder (140 g), with addition of 7 L of H_2O, was placed into 70 °C hot water bath for 2 h for the polysaccharide extraction. The extract liquid was then taken into tubes, centrifuged at 4400 rcf/g for 10 min, and kept at 20 °C for 15 min. The mycelia precipitate was separated from the crude water-soluble polysaccharide supernatant. The supernatant was then concentrated in a rotary evaporator under reduced pressure at 60 °C to get 850 mL of vacuum-concentrated liquid. The concentrated extract solution was precipitated with 3.4 L ethanol and kept at 4 °C overnight. The precipitate was obtained by centrifugation at 4400 rcf/g for 15 min and then dried at 45 °C for 2 h, giving the crude polysaccharides. The crude polysaccharide was then re-dissolved with 800 ml H_2O. The re-dissolved polysaccharide (500 mL) was treated with Sevag reagent (*n*-butanol: chloroform = 1:4, v/v, 120 mL) to remove the proteins inside the solution. The mixture was violently oscillated for 30 min and centrifugated to remove the denatured proteins at the interface between water layer and Sevag reagent layer. The above operation was repeated until no denatured proteins appeared. In order to decolor the solution, 1.5% (v/v) activated charcoal was added to the Sevag-treated crude polysaccharide, with thermostatic water bathing for 40 min, then the polysaccharide solution was poured into a dialysis bag, with both ends tightened up, and placed into H_2O. The water was changed every 4 h, until the color of the dialysate did not change.

2. **Preparation of Freeze-Dried Polysaccharide Samples from *Ganoderma* Genus**

Each liquid sample obtained from the steps mentioned above was pipetted and placed into a petri dish at −60 °C for 48 h to freeze for drying (FD-1D-50, Bilon, China). Then we obtained the following samples:

(a) The GL powder means the drying powder from wet *Ganoderma* culture mycelia.
(b) The crude GLPS means the extracted crude polysaccharide from the dried powder in 70 °C of hot water.
(c) The GLPS after condensing is the condensing supernatant using rotary vacuum approach.
(d) The GLPS after ethanol precipitating means the condensed remnant after ethanol precipitating process.
(e) The GLPS after Sevag means the polysaccharides removing proteins with the by Sevag method.
(f) The GLPS after dialysis means the polysaccharide samples which remove small molecular impurity substances after dialyzing operation. These samples were then examined by infrared spectroscopy.

3. **Measurement of Polysaccharides in Dried Mycelia of Different *Ganoderma* Stains**

Solutions prepared were 2 mL of 0.012, 0.024, 0.036, 0.048, 0.06, 0.072, and 0.084 mg/mL glucose solutions. Then, 6 mL of anthrone reagent was added to each glucose solution, and the solution was first kept at room temperature for 15 min and then stored on ice for 15 min. When the test tubes were cooled, 3 mL of each sample was read at 625 nm wavelength using UV–vis spectrophotometer (Shimadzu UV-2550, Japan). The spectrum value at 625 nm was recorded for analysis. A standard curve for total carbohydrate assay was generated. The determination coefficient (R_2) of glucose standard curve is 0.9903, with the standard error less than 0.001. The lyophilized sample (0.1 g) was mixed with 10 mL of H_2O and placed steady for 1 h. After that, the mixture was placed in 70 °C hot water bath for 2 h, and centrifugated after cooling, and the precipitate was discarded. The supernatant was diluted 20 times and sample solution (2 mL) was pipetted into a test tube for measurement. Sulfuric acid solution (6 mL) was added into the test tube and mixed with the sample together. The mixed sample was measured at 625 nm using the UV–vis spectrophotometer. The content of polysaccharide was then calculated referring to standard curve above (g glucose/100 g sample). And the *Ganoderma* polysaccharide content was used as reference value for the quantification model.

4. **Measurements of mid-IR Spectra**

The samples for mid-IR measurement were prepared by mixing 2 mg of freeze-dried *Ganoderma* mycelia samples with 200 mg of dried potassium bromide followed by pressing under pressure 15 MPa for 3 min to make a disk pellet. The samples were then subjected to mid-IR measurements, and the spectral range (4000~400 cm^{-1}) was recorded using a Bruker ALPHA-T instrument (Bruker Optik GmbH, Ettlingen, Germany) with a resolution of 4 cm^{-1} and 64 scans per sample. The results were then analyzed using OPUS 7.0 data processing software.

5. **Measurement of NIR Spectra**

A FT-NIR spectrometer (NIR MPA, Bruker Optik GmbH, Germany) was used to collect the diffuse reflection spectra, with a resolution of 16 cm^{-1} and 32 scans per sample ranged from 12,500 to 4000 cm^{-1}. Each sample was tested several times for the average. These results were then analyzed by OPUS 7.0 data processing software.

5.5 The Analytical Methods for Quantitative Analysis of Nucleosides or Nucleobases in *Ganoderma*

Up to now, only three papers reported the quantitative analysis of nucleosides or nucleobases in *Ganoderma* [8–10]. A total of 18 nucleosides or nucleobases in *Ganoderma* were quantitatively analyzed in 14 batches of pileus, stipe, or the whole mushroom of *Ganoderma*, including *G. lucidum*, *G. sinense*, and *G. atrum*, respectively [8–10]. Three methods, including HPLC-DAD-MS (reverse phase column), CE-MS, and ZIC-HILIC (with DAD detector), were used in analysis, respectively [8–10]. Compared to the traditional reverse phase column, ZIC-HILIC separation with low aqueous and high organic mobile phase has been demonstrated as a valuable complementary method to HPLC-DAD-MS (reverse phase column). In addition, ZIC-HILIC separation is ideal for ESI detection with high sensitivity due to low water percentage [9]. The flow rate in CE is in the few nanoliters/minute range, about 1/10 of nano-HPLC, and ESI technique just needs the same amount of sample injection, so CE peaks provide higher S/N, partially compensating for its disadvantage of low concentration detection limit [10].

5.5.1 *Quantitative Analysis of Nucleosides or Nucleobases in* **Ganoderma** *Using HPLC-DAD-MS (Reverse Phase Column) [8]*

An HPLC-DAD-MS method was developed to quantitatively analyze two species of *Ganoderma*, the dried sporophores of *G. lucidum* and *G. sinense*, including pileus and stipe. A total of 11 compounds, including six nucleosides (adenosine, cytidine, guanosine, inosine, thymidine, and uridine) and five nucleobases (adenine, guanine, hypoxanthine, thymine, and uracil), were quantitatively determined.

Quantitative analyses were performed on an Agilent 1100 series LC/MSD VL trap system (Agilent, USA) equipped with an autosampler, a high-voltage power supply, a diode array detector, an ESI source, and an ion trap analyzer. An Agilent AORBAX Eclipse XDB C18 column (3.5 μm, 4.6 mm × 150 mm) with an XDB C8 guard column (5.0 μm, 3.9 mm × 20 mm) was used.

The separation was achieved using gradient elution with 5 mM ammonium acetate solution and methanol: 0~5% methanol in 0~10 min and 5~20% methanol in 10~30 min. The flow rate was set at 0.5 mL/min. The column oven is at 25 °C. The DAD detection wavelength was set at 254 nm. An aliquot of 10 μL solution obtained from above was injected for HPLC analysis. The MS spectra were acquired in positive ion mode. The full-scan mass spectra were obtained from *m/z* 50 to 400.

Calibration curves were established for quantitative analysis. Compound solutions were prepared in methanol-water (1:1). After appropriate dilution, working solutions of various concentrations were freshly prepared.

One gram of pulverized pileus or stipe of the fruiting body of *Ganoderma* was extracted with 15 mL of water by ultrasonic technology for 45 min at room temperature. The supernatant was collected and concentrated to about 1 mL under a stream of nitrogen. Then the solution was ready for HPLC analysis.

The contents of nine nucleosides and nucleobases in pileus and stipe of six *Ganoderma* fruiting bodies were shown in Table 5.4. Uridine was the most abundant nucleoside among 11 detected compounds in these *Ganoderma* fruiting bodies, and the amounts of the nucleosides and nucleobases in *G. lucidum* were more abundant than those in *G. sinense*.

5.5.2 *Quantitative Analysis of Nucleosides or Nucleobases in* **Ganoderma** *Using HPLC-DAD-MS (ZIC-HILIC) [9]*

A ZIC-HILLC method was developed for the simultaneous determination of 16 nucleosides (adenosine, 2-deoxyadenosine, cytidine, 2-doxycytidine, guanosine, 2-deoxyguanosine, inosine, thymidine, uridine, and 2-deoxyuridine) and nucleobases (adenine, 2-deoxyadenosine, cytosine, guanine, thymine, and uracil).

Analysis was performed on an Agilent Series 1200 liquid chromatograph, equipped with a vacuum degasser, a quaternary pump, an autosampler, and a

Table 5.4 The contents of nine nucleosides and nucleobases in pileus and stipe of six *Ganoderma* fruiting bodies (μg/g)

Samples	*G. lucidum* (Zhejiang)		*G. lucidum* (Shandong)		*G. lucidum* (Guangxi)		*G. lucidum* (Sichuan)		*G. sinense* (Xinjiang)		*G. sinense* (Sichuan)	
Analytes (μg/g)	Pileus	Stipe	Pileus	Stipe	Pileus	Stipe	Pileus	Stipe	Pileus	Stipe	Pileus	Stipe
Uracil	32.52	9.41	67.47	22.69	7.38	–	73.17	31.88	–	–	62.96	37.99
Cytidine	94.98	48.06	117.78	43.11	39.19	–	200.57	50.70	3.04	8.67	15.04	6.22
Hypoxanthine	110.83	47.74	30.40	13.64	52.77	–	38.34	13.96	1.33	2.54	25.37	12.80
Uridine	264.93	135.57	252.83	117.14	185.22	–	310.71	113.91	67.06	61.49	118.43	102.65
Thymine	7.89	–	4.67	–	5.81	–	17.19	–	–	–	11.88	3.62
Inosine	4.54	13.47	96.06	27.48	–	–	58.03	16.66	7.25	8.28	17.04	1.83
Guanosine	4.18	18.63	174.91	63.34	–	11.44	214.62	39.51	10.81	10.64	33.62	17.21
Thymidine	34.46	19.26	25.10	17.14	9.60	10.37	36.02	27.57	6.62	5.51	–	–
Adenosine	–	4.11	97.89	29.35	3.85	–	269.03	17.34	–	1.76	22.19	10.16
Total	554.31	326.23	867.11	333.89	303.81	21.81	1217.68	311.52	96.11	98.89	306.53	192.48

diode-array detector (DAD). A Merck ZIC HILIC column (3.5 μm, 100 × 4.6 mm) and a guard column (20 × 2.1 mm) were used for separation.

The mobile phase was composed of solvent A (CAN modified with 3 mM ammonium acetate) and solvent B (pH 6; 15 mM ammonium acetate in water). The gradient procedure was as follows: 0~18 min, 3~5% B; 18~19 min, 5~10% B; and 19~30 min, 10~20% B. The flow rate was 1.0 mL/min. The temperature of column compartment was kept at 25 °C, and the peaks were monitored at a wavelength of 254 nm.

Calibration curves were established for quantitative analysis. Standard solutions were made in acetonitrile-water (1:1). The desired concentration of the solvent was freshly made after dilution.

A total of eight natural or cultivated *Ganoderma* materials (1 g) were mixed with 20 mL of water, using ultrasonic extraction at room temperature for 10 min. The supernatant was ready for injection into HPLC.

A summary of the nucleobase and nucleoside analytes from these samples were shown in Table 5.5. The total amounts of the nucleobase and nucleoside in *G. lucidum* were higher than those in *G. sinense*, which is in accordance with the reported data. The contents of sixteen nucleosides and nucleobases in seven batches of *Ganoderma* (μg/g) were shown in Table 5.5.

5.5.3 Quantitative Analysis of Nucleosides or Nucleobases in Ganoderma *Using CE-MS [10]*

The eight nucleosides or nucleobases were separated by capillary electrophoresis (CE) with ESI-MS detector in an optimized electrolyte solution made of 10% MeOH containing 100 mmol/L formic acid. The column used for separation (kept at 25 °C) was fused silica capillary (120 cm × 50 μm). The applied voltage was 25 kV. The ESI-MS analysis was performed in positive ion mode, and the full-scan range was from *m/z* 50 to 350. The contents of eight nucleosides and nucleobases in seven batches of *G. lucidum* and *G. sinense* (μg/g) were shown in Table 5.6.

5.6 The Analytical Methods for Quantitative Analysis of Ergothioneine in *Ganoderma* [11]

Ergothioneine in *Ganoderma* were determined by HPLC-UV. The samples were analyzed by kromasil 100-5 NH_2 column with elution of acetonitrile 5 mmol/L ammonium acetate (80:20, v/v) as a mobile phase. The detection wavelength was 254 nm. The content of ergothioneine in *G. lucidum* is 0.064%.

Table 5.5 The contents of sixteen nucleosides and nucleobases in eight batches of *Ganoderma* fruiting bodies ($\mu g/g$)

Samples	*G. atrum*		*G. lucidum*			*G. sinense*		
Analytes (μg/g)	Jiangxi	Zhejiang	Anhui	Shandong	Hubei	Natural-1	Natural-2	Fujian
Adenosine	39.76	26.32	8.17	17.14	53.87	–	25.67	7.40
2-deoxyadenosine	3.36	–	–	–	–	6.36	2.25	1.29
Cytidine	–	–	–	–	–	–	5.39	2.91
2-deoxycytidine	6.05	12.36	23.30	9.99	–	–	7.91	13.97
Guanosine	44.26	–	–	21.03	–	–	45.65	6.38
2-deoxyguanosine	10.23	–	–	–	–	–	24.78	18.49
Inosine	3.25	13.48	27.55	42.65	–	–	13.23	–
Thymidine	–	138.23	117.90	260.22	474.44	235.96	23.77	15.57
Uridine	3.64	–	–	8.40	–	41.42	4.36	–
2-deoxyuridine	–	25.15	28.75	52.83	173.05	105.66	6.10	10.99
Adenine	–	56.54	46.80	67.65	147.29	202.80	12.27	11.25
2-deoxyadenosine	10.23	–	–	–	–	–	24.78	18.49
Cytosine	10.94	8.31	43.39	19.65	–	88.41	34.00	34.01
Guanine	44.26	–	–	21.03	–	–	45.65	6.38
Thymine	–	138.23	117.90	260.22	474.44	235.96	23.77	15.57
Uracil	1.26	145.40	62.87	210.65	406.12	645.43	25.91	30.67
Total	127.11	430.69	362.01	724.21	1270.24	1333.11	242.66	154.53

Table 5.6 The contents of eight nucleosides and nucleobases in seven batches of *G. lucidum* and *G. sinense* (μg/g)

Samples	*G. lucidum*				*G. sinense*		
Analytes (μg/g)	Anhui	Sichun	Shandong	Yunnan	Anhui	Sichuan	Shandong
Cytidine	35.08	148.50	110.72	289.15	64.30	56.54	81.58
Adenosine	115.87	559.93	428.57	859.02	195.97	152.01	344.31
Hypoxanthine	–	23.60	25.18	26.29	–	–	–
Guanosine	–	262.42	120.21	502.16	189.42	102.59	248.85
Inosine	–	61.00	45.24	2310.50	–	–	–
Uridine	–	324.24	–	446.94	375.42	341.75	335.40
Cytosine	14.57	36.52	36.20	41.98	12.91	–	47.37
Guanine	15.05	–	–	–	–	–	–
Total	180.57	1416.21	766.12	4476.04	838.02	652.89	1057.51

References

1. USP38–NF33 (2014) United Book Press, Baltimore, pp 9033
2. Gao J, Sato N, Hattori M, Ma C (2013) The simultaneous quantification of *Ganoderma* acids and alcohols using ultra high-performance liquid chromatography-mass spectrometry in dynamic selected reaction monitoring mode. J Pharm Biomed Anal 74:246–249
3. Krystyna S, Janusz S, Renata L, Marek S, Krzysztof S, Kazimierz G, Anna M (2012) Evaluation of polysaccharides content in fruit bodies and their antimicrobial activity of four *Ganoderma lucidum* (W Curt.: Fr.) P. Karst. strains cultivated on different wood type substrates. Acta Soc Bot Pol 81(1):17–21
4. Ma Y, He H, Wu J, Wang C, Chao K, Huang Q (2018) Assessment of polysaccharides from mycelia of genus *Ganoderma* by mid-infrared and near-infrared spectroscopy. Sci Rep 8(1):10
5. Chen Y, Xie M, Zhang H, Wang Y, Nie S, Li C (2012) Quantification of total polysaccharides and triterpenoids in *Ganoderma lucidum* and *Ganoderma atrum* by near infrared spectroscopy and chemometrics. Food Chem 135:268–275
6. Zhao J, Zhang X, Li S, Yang F, Wang Y, Ye W (2006) Quality evaluation of *Ganoderma* through simultaneous determination of nine triterpenes and sterols using pressurized liquid extraction and high performance liquid chromatography. J Sep Sci 29:2609–2615
7. Ha D, Loan L, Hung T, Han L, Khoi N, Dung L, Min B, Nguyen N (2015) An improved HPLC-DAD method for quantitative comparisons of triterpenes in *Ganoderma lucidum* and its five related species originating from Vietnam. Molecules 20:1059–1077
8. Gao J, Leung K, Wang Y, Lai C, Li S, Hu L, Lu G, Jiang Z, Yu Z (2007) Qualitative and quantitative analyses of nucleosides and nucleobases in *Ganoderma* spp. by HPLC-DAD-MS. J Pharm Biomed Anal 44:807–811
9. Chen Y, Bicker W, Wu J, Xie M, Lindner W (2012) Simultaneous determination of 16 nucleosides and nucleobases by hydrophilic interaction chromatography and its application to the quality evaluation of *Ganoderma*. J Agric Food Chem 60:4243–4252
10. Yang F, Zhang X, Ge L, Tan S (2011) Determination of nucleosides and related compounds in *Ganoderma* by CE-MS. J China Pharm Univ 42(4):337–341
11. Zhao Y, Lei Z, Liu C, Liu D (2016) Determination of ergothioneine in multiple species of fungi by HPLC. ShipinYanjiu Yu Kaifa 37:117–119
12. Gao J, Nakamura N, Min B, Hirakawa A, Zuo F, Hattori M (2004) Quantitative determination of bitter principles in specimens of *Ganoderma lucidum* using high-performance liquid chromatography and its application to the evaluation of *Ganoderma* products. Chem Pharm Bull 52(6):688–695

13. Liu W, Zhang J, Han W, Liu Y, Feng J, Tang C, Feng N (2004) One single standard substance for the simultaneous determination of 17 triterpenses in *Ganoderma lingzhi* and its related species using high-performance liquid chromagraphy. J Chromatogr B 52(6):688–695
14. Liu J, Kurashikia K, Fukuta A, Kaneko S, Suimic Y, Shimizu K, Kondo R (2012) Quantitative determination of the representative triterpenoids in the extracts of *Ganoderma lucidum* with different growth stages using high-performance liquid chromatography for evaluation of their 5α-reductase inhibitory properties. Food Chem 133:1034–1038
15. Wang X, Yang M, Guan S, Liu R, Xia J, Bi K, Guo D (2006) Quantitative determination of six major triterpenoids in *Ganoderma lucidum* and related species by high performance liquid chromatography. J Pharm Biomed Anal 41:838–844
16. Li B, Liu C, Wang H, Chen R (2007) The study of quantitative analysis of total triterpenes in *Ganoderma*. Chin J Chin Mater Med 32(12):1234–1236
17. Li B, Gu H, Li Y, Chao L, Wang H, Kang J, Wu C, Chen R (2007) Determination of nine triterpenoid acids from *Ganoderma lucidum* of different producing area by HPLC. Chin J Chin Mater Med 32(12):1234–1236

Chapter 6
Researches and Application of Ganoderma Spores Powder

Jianhua Xu and Peng Li

Abstract *Ganoderma lucidum* spores (GLS) are the mature germ cells of *Ganoderma lucidum*. They have all the genetic substances and similar active components of *Ganoderma lucidum*. Similar to the fruiting body of *Ganoderma lucidum*, ganoderma spores powder has the effect of regulating immunity, antitumor, antioxidation, and protecting cells and so on. In recent decades, with the development of the technology of breaking the wall of *Ganoderma lucidum* spores and the technology of extracting and preparing, the researches and application of *Ganoderma lucidum* spores powder have made great progress.

Keywords Ganoderma · Lingzhi · Spore · Polysaccharides · Triterpenes · Immune · Tumor

Ganoderma was published in *Shen Nong Ben Cao Jing* (*Shennong's herbal classic*) and listed as effective and nontoxic drug that have been used in traditional Chinese medicine for thousands of years. Modern research shows that polysaccharides, proteins, and triterpenoids obtained from *Ganoderma lucidum* (*G. lucidum*) are the key compounds in this fungus that are responsible for its main biological effects. However, it has been recently reported that these biomolecules can also be obtained from *G. lucidum* spores (GLSs). This is in part because of the recent development of methods for breaking the spore wall and phytochemical technique. Therefore in recent decades *G. lucidum* spores caused special attention to scientists. It has been verified that GLS has multiple functions similar to the fruiting bodies of *G. lucidum*, such as immune regulation, antitumor, neuroprotection, improving liver function, protection from radiotherapy- and chemotherapy-induced injuries, antioxidation and free radical scavenging effect, etc. The research progress and clinical application of *G. lucidum* spores powder are reviewed in the present article.

J. Xu (✉) · P. Li
Department of Pharmacology, School of Pharmacy, Fujian Provincial Key Laboratory of Natural Medicine Pharmacology, Fujian Medical University, Fuzhou, Fujian, China
e-mail: xjh@fjmu.edu.cn

Z. Lin, B. Yang (eds.), *Ganoderma and Health*, Advances in Experimental Medicine and Biology 1181, https://doi.org/10.1007/978-981-13-9867-4_6

Fig. 6.1 The photographs of *G. lucidum* spores. (**a**) Diagram of *G. lucidum* spores deposited on the upper side of Ganoderma pilei. (**b**) Diagram of sporoderm-broken and sporoderm-unbroken *Ganoderma lucidum* spores powder. (Photos provided by Fujian Xianzhilou Biological Science & Technology Co., Ltd. of China)

6.1 Spore Structure of *Ganoderma lucidum*

Ganoderma lucidum spores are the germ cells of *Ganoderma lucidum*, which are extremely small particles ejected from the fungus cap of *Ganoderma lucidum* at maturity (Fig. 6.1). They are brown, ovate tiny particles wrapped by outer bilayers of sporoderm about 8.5~11.2 μm × 5.2~6.9 μm in size (Fig. 6.2a, b) [1, 2]. The spore walls are bitunicate with pillars contacted among exosporium and endosporium, with a thickness of 0.8~1.1 μm (Fig. 6.2c, d). Some inorganic elements such as magnesium, aluminum, silicon, phosphate, sulfur, chlorine, potassium, calcium, iron, and nickel were detected in the sporoderm of *G. lucidum*. Silicon (19.01%) and calcium (24.31%) were detected at higher levels [3]. The sporoderm is so extremely hard and resilient that the bioactive substances inside the spores are difficult to be released.

6.2 Collection and Processing of *Ganoderma lucidum* Spores Powder

6.2.1 Collection of Spores Powder [4]

Fifteen days or so after the disappearance of the yellow-white growth ring around the pileus of *G. lucidum*, when the spores are about to eject, the spores can be collected. Ambient temperature at the time of collection is advisable at 25–33 °C. The methods of the collection are as follows:

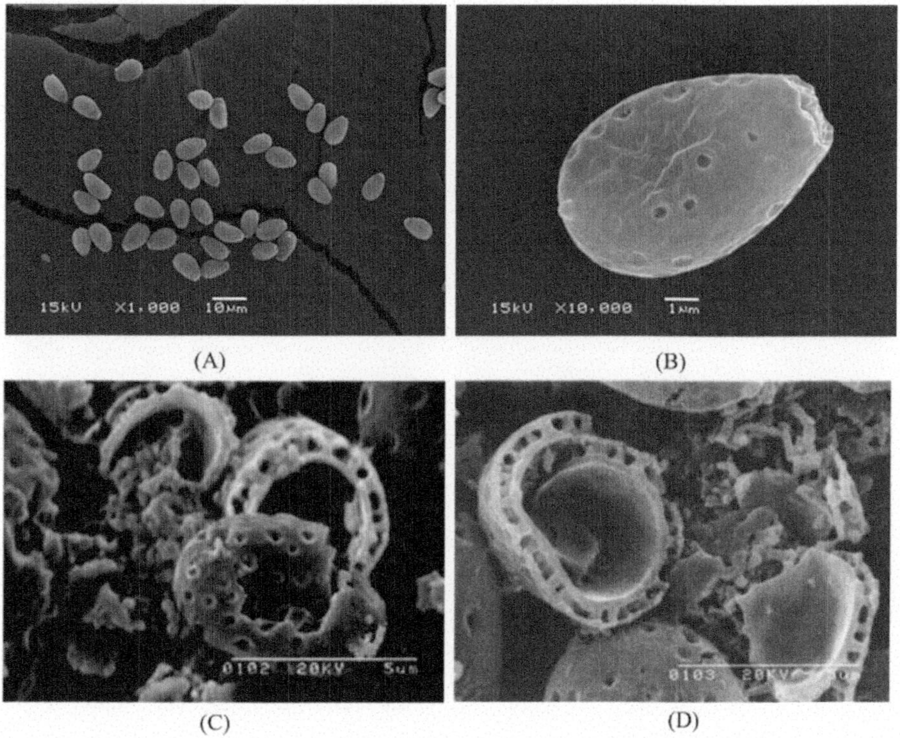

Fig. 6.2 Electronic microscopic photographs illustrating the microscopic structures of *G. lucidum* spores. (**a**) Diagram of ovoid *G. lucidum* spores. (**b**) Locally amplificatory diagram for one *G. lucidum* spore. (**c**) Section diagram of *G. lucidum* spores. (**d**) Section diagram showing the bilayer episporium of one *G. lucidum* spore [2]

6.2.1.1 Kraft-Paper Sheath Collection Method

When the yellow-white growth ring at the margin of the pileus is about to disappear, a thin film is laid on the ground to isolate the soil. Another suitable size clean plastic film is laid on the base of the stipe of *G. lucidum*. A white card paper or glossy paper was rolled into a cylinder and placed on the film. The diameter of the cylinder should be 2–4 cm larger than the pileus of *G. lucidum*. After the cylinder entangles the *G. lucidum*, it is covered with square cardboard to prevent the spores powder from escaping. The gap between the cover plate and the pileus of *G. lucidum* should be 5 cm (Fig. 6.3). Make small holes in the cylinder to increase air permeability and maintain air relative humidity at 75–80%.

After removing the cylinder, the spores powder on the inside of the cylinder and the pileus of *G. lucidum* are first collected. Then the spores powder on the film are removed by spoon. *G. lucidum* spores are collected every 15 days until 45 days.

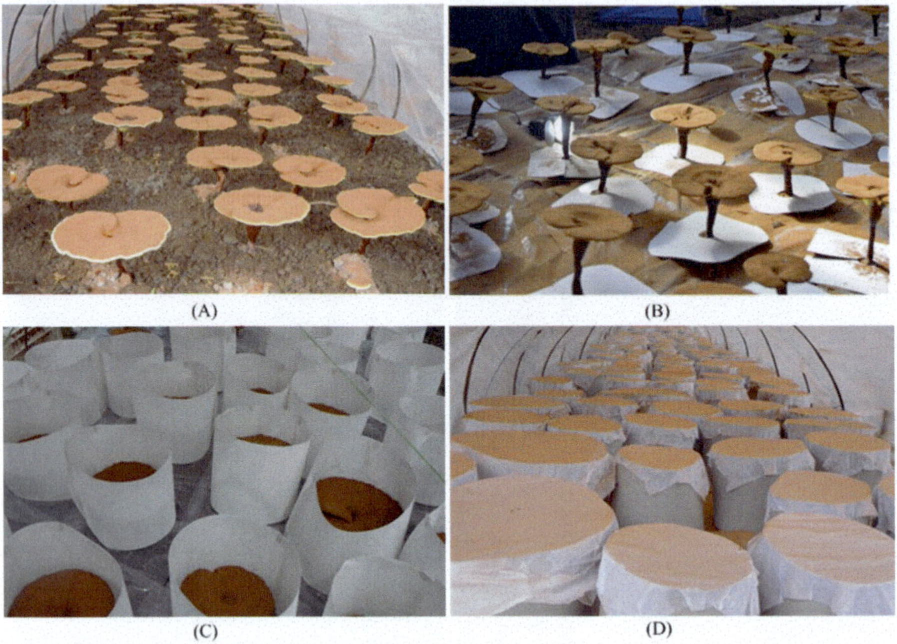

Fig. 6.3 The collecting process of *G. lucidum* spores. (**a**) Diagram of *G. lucidum* grown in a greenhouse. (**b**) Diagram showing a thin film on the base of *G. lucidum*. (**c**) Diagram showing a kraft-paper sheath around *G. lucidum*. (**d**) Diagram showing a kraft-paper sheath with a cover (Photos provided by Fujian Xianzhilou Biological Science and Technology Co., Ltd. of China)

6.2.1.2 Vacuum Suction Methods

Spores powder are collected by a spore collector consisting of a fan and a cloth bag. When the *G. lucidum* spores begin to release, the spores collector is placed 1~1.5 m above the ground in the middle of the shed where *G. lucidum* is growing. Then the fan is launched to form negative pressure to collect spores powder.

6.2.1.3 Plastic Film Mulching Methods

Two layers of strip membrane are laid in the middle of every two rows of mature *G. lucidum* to collect the fallen *G. lucidum* spores powder. The spores on the surface of the pileus of *G. lucidum* are brushed into a special container with a special soft brush, and then the spores powder on the membrane are collected. When collecting, only the spores powder on the upper layer of membrane are collected, and the spores powder on the lower layer are discarded.

6.2.2 Spore's Wall Breaking and Its Efficacy

It has been reported that *G. lucidum* spores could be administered in a dormant or germinative phase. Germinating spores had higher bioactivity than dormant spores, suggesting that these effects depend on whether the sporoderm has been broken or not [5]. The method breaking the sporoderm of *G. lucidum* spores can be physical, physicochemical, enzymatic, and zymotic (Table 6.1). Some studies have shown that the amount of bioactive components extracted from *G. lucidum* spores depend on if the sporoderm has been broken. After breaking down the spore wall, the yield of aqueous polysaccharides was increased by 40.08%, and a greater biological effect were also observed [14]. The percentage of carbohydrate content of sporoderm-broken *G. lucidum* spores (1.57% ± 0.06%) was higher than that of sporoderm-unbroken spores (0.41% ± 0.01%) [7]. The extraction yield of polysaccharides was increased from 0.94% using whole spores to 2.98% using sporoderm-broken spores [10].

6.2.3 Chemical Components

The chemical components of *G. lucidum* spores contain triterpenes, polysaccharides, amino acid peptides, sterols, alkaloids, fatty acids, vitamins, inorganic ions, and so on. Studies have shown that lanostane-type triterpenes are one of the most important bioactive components of *G. lucidum* spores. To our knowledge, about 29 different types of triterpenes have been isolated and identified from *G. lucidum* spores. These triterpenes include ganosporeric acid A; ganoderic acid B, C, and E;

Table 6.1 Methods breaking the sporoderm of *Ganoderma Lucidum* spores

Method	Broken efficacy (%)	References
Physical		
Ultrasonography	60–80	[1]
Grinding	NR	[6]
Ultrafine grinding	100	[7]
High-speed centrifugal shearing pulverizer	100	[8]
Physical/chemical		
Chemical reactions cooperated with sonication and refrigeration	NR	[9]
Supercritical extraction with CO_2	NR	[10]
Microwave and alcohol	NR	[11]
Enzymatic		
Inoculating spores with *G. lucidum* mycelia which can release different types of enzymes	NR	[6, 12]
Not specified	99.8	[5]
Zymotic		
Fermentation with *Lactobacillus plantarum*	NR	[13]

ganodermanontriol [15]; ganosporelactone A and ganosporelactone B [16], methyl ganoderate A and methyl ganoderate B [17]; ganoderic acid H [18]; ganoderic acid γ, δ, ε, ζ, η, and θ, ganolucidic acid D, and ganoderic acid C2 [19]; lucidumol A, ganoderic acid β, ganodermanondiol, ganoderiol F, ganoderic acid A, ganoderic acid C1, and ganolucidic acid A [20], and lucidenic acid SP1, ganoderic acid C6, ganoderic acid G, and lucidumol B [21]. Of these compounds, ganosporelactone A, ganosporelactone B, and lucidenic acid SP1 were isolated and identified only in the spores of *G. lucidum*, which had not been reported from the fruiting body of *G. lucidum*.

Some polysaccharides with different biological activities have been isolated from *G. lucidum* spores. Most polysaccharides are glucans with different combinations of glucose, including linear (1→3)-α-D-glucan [22], branched α-D-glucan with (1→4)-α-D-glucan backbone [23], β-D-glucan with (1→3)-β-D-glucan backbone [24], β-D-glucan with (1→6)- β-D-glucan backbone [25], mixed-linkage α-β-D-glucans [26, 27], and β-D-glucan with a mixed (1→3), (1→4), (1→6)-linked β-D-glucan backbone [28].

Using amino acid automatic analyzer, 18 kinds of common amino acids have been detected from *G. lucidum* spores. The content of total amino acids is 7.29–7.71 mg/100 mg, among which the content of methionine is as high as 3.30–3.48 mg/100 mg. The content of total essential amino acids in human body accounts for 69.4–70.4% of the total amino acids. These 18 common amino acids are aspartic acid, methionine, threonine, isoleucine, serine, leucine, glutamic acid, tyrosine, proline, phenylalanine, glycine, lysine, alanine, tryptophan, cystine, histidine, valine, and arginine [29].

Vitamins are mainly vitamin E, which is more than 60 mg per gram, and also contain a small amount of vitamin C, about 5.6 mg per gram [30].

At present, about 18 sterols have been isolated from *G. lucidum* spores, including ergosta-7,22-dien-3β,5α,6β-triol, ergosta-7,22-dien-3β,5α,6α-triol, ergosta-7,9,22-trien-3β,5α,6α- triol, ergosterol palmitate, ergosta-4,6,8(14), 22-tetraen-3-one, ergosterol [31], ergosterol peroxide, ergosta-7,22-diene-3β-yl pentadecanoate [32], ganodermasides A, ganodermasides B [33], ganodermasides C, ganodermasides D [34], 22E, 24R-ergosta-7,22-diene-3β,5α,6β,9α,14α-pentol, 22E, 24R-ergosta-7,22-diene-3β,5α,6β,9α,-tetraol, 3,5-dihydroxy-(22E, 24R)-ergosta-7,22-diene-6-one, 3β,5α,9α-trihydroxy-(22E, 24R)-ergosta-7,22-diene-6-one [18], ganoderin A, and stellasterol [35].

The alkaloids in *G. lucidum* mainly include choline and betaine [36].

Gas-liquid chromatography of spore lipids obtained spores with a broken sporoderm detected 15 kinds of fatty acids, the main compounds being palmitic acid (6.12%), stearic acid (4.97%), oleic acid (67.11%), and linoleic acid (9.63%) [37].

The inorganic elements in *G. lucidum* spores include calcium, phosphorus, iron, magnesium, sodium, zinc, germanium, cobalt, chromium, selenium, and so on [29].

6.2.4 *Quality Study and Quality Control*

The quality of *Ganoderma lucidum* spores powder can be evaluated comprehensively from the aspects of wall breaking rate, oil content, active ingredients, heavy metals, and solvent residue.

6.2.4.1 Identification

Thin-layer chromatography (TLC), UV spectrophotometry, and ^{1}H-NMR spectrophotometry could be used to distinguish the *G. lucidum* fine powder, *G. lucidum* spores powder, and sporoderm-broken *G. lucidum* spores powder. Jin Y (2006) used TLC to identify sporoderm-broken *G. lucidum* spores powder capsules, using the top solution of petroleum ether-ethyl formate-formic acid (15:5:1) as developing solvent. The method had the advantages of simple operation procedure and short separation time and could distinguish the sporoderm-broken *G. lucidum* spores powder from the *G. lucidum* fine powder [38]. Wang X et al. (2008) used UV spectrophotometry to identify *G. lucidum* spores powder and sporoderm-broken *G. lucidum* spores powder. The *G. lucidum* spores powder and sporoderm-broken *G. lucidum* spores powder were extracted by microwave with different solvents, and the same solvent extract was detected by UV spectrophotometry. The results showed that the UV spectrum of water extract of sporoderm-broken *G. lucidum* spores powder was different from that of *G. lucidum* spores powder, which could be used to distinguish *G. lucidum* spores from sporoderm-broken *G. lucidum* spores [39]. Wu YK et al. (2008) used ^{1}H-NMR spectrum to identify spores powder and sporoderm-broken spores powder from *G. lucidum*. Spores powder and sporoderm-broken spores powder were extracted by microwave with different solvents, and the extracts were studied by ^{1}H-NMR, so as to achieve ^{1}H-NMR spectrum analysis of these two spores powder. The ^{1}H-NMR spectra of chloroform extracts were basically the same, but the ^{1}H-NMR spectra of water extracts were different, which could be used for the differentiation between spores and sporoderm-broken spores from *G. lucidum* [40].

6.2.4.2 The Sporoderm-Broken Rate

The detection methods for the sporoderm-broken rate of *G. lucidum* spores mainly include counting method [41], water film loading in combination with microscopy [42], and suspension in combination with physical method [43]. The sporoderm-broken rate of Ganoderma spores powder shall not be less than 95%.

6.2.4.3 Oil Content

G. lucidum spores oil is one of the main active substances in spores powder. It has a wide range of pharmacological activities such as inhibiting tumor cell growth, anti-inflammation, and liver protection. The oil content of *G. lucidum* spores directly reflects its inherent quality. Yao ZH et al. (2007) determined the oil content of sporoderm-broken *G. lucidum* spores powder by gravimetric method. According to the test results, the oil content of sporoderm-broken *G. lucidum* spores powder should not be lower than 20% [42].

6.2.4.4 The Determination of Triterpenes and Polysaccharides

Triterpenoids are one of the main active substances in *G. lucidum* spores powder, which have the functions of detoxification, liver protection, antitumor, and anti-inflammatory. The chemical analysis of triterpenes is usually carried out by ultraviolet spectrophotometry and HPLC. Chen GZ et al. (2011) determined the content of total triterpenes in sporoderm-broken *G. lucidum* spores powder by UV spectrophotometry with ursolic acid as reference substance, water bath at 60 °C for 15 min, 548 nm as the measured wavelength. The color reagent was 5% vanillin-glacial acetic acid and perchloric acid. This method was simple, rapid, stable, and reproducible. It could be used for quality control of sporoderm-broken *G. lucidum* spores powder. The total triterpenes content expressed as ursolic acid equivalents shall be not less than 1.0% [44]. Ganoderic acid A, ganoderic acid G, and ganoderic acid C_2 were the three main triterpenoid acids in *G. lucidum*. A HPLC method for quantitative analysis of these three ganoderic acids was carried out for overall quality evaluation of *G. lucidum* spores. The HPLC method was performed on Diamonsil C_8 column (250 mm × 4.6 mm, 5 μm), with acetonitrile-0.03% phosphoric acid (28:72) as the mobile phase. The content of triterpenoid acids in the fruiting body of *G. lucidum* was significantly higher than that of *G. lucidum* spores. The content of ganoderma acids in commercially available spores powder was lower and the difference was greater. The method was proved to be convenient and reliable, so it could be used for the quantitative analysis of ganoderic acids in *G. lucidum* spores [45].

Polysaccharides are one of the main components of *G. lucidum* spores powder. Modern pharmacological studies have found that it can enhance immunity, antitumor, antioxidative, and anti-aging. The chemical analysis of polysaccharides is usually carried out by UV-vis spectrophotometry [46], HPLC [47], ion chromatography [48], gas chromatography-mass spectrometry, and gel permeation chromatography [49]. The processing standard of traditional Chinese medicine decoction pieces in Fujian province of China (2012) stipulates that the polysaccharide content expressed as glucose equivalents in sporoderm-broken ganoderma spores powder shall be not less than 0.9%.

6.2.4.5 Heavy Metal Content Measurement

Wang WY et al. (2014) used atomic fluorescence spectrometry and atomic absorption spectrophotometry to determine the content of As, Hg, Pb, and Cd in *G. lucidum* spores powder. The highest contents of As, Hg, Pb, and Cd in *G. lucidum* spores powder were 0.342, 0.043, 10.495, and 0.649 mg/kg, respectively. This method is sensitive, efficient, and accurate [50]. Gao LH et al. (2008) used hydride generation-atomic fluorescence method to determine As in *G. lucidum* spores powder, which was time saving, reagent saving, and highly sensitive. It is suitable for the determination of As in *G. lucidum* spores powder [51]. Li J et al. (2014) used ICP-MS instrument to determine the content of Pb, As, and Hg heavy metal elements in *G. lucidum* spores powder. This method is simple, rapid, efficient, and accurate and can objectively evaluate the effects of environmental pollution on *G. lucidum* spores powder [52].

In the process of *G. lucidum* spores breaking the wall, there may be residual solvents which can affect the quality and application safety of *G. lucidum* spores powder. Therefore, it is necessary to detect the residual solvents. Gao WC et al. (2014) established a direct injection gas chromatography method to determine the residual amounts of eight organic solvents in health food of sporoderm-broken *G. lucidum* spores powder, including methanol, ethanol, acetone, n-hexane, ethyl acetate, benzene, n-butanol, and divinyl benzene. This method is simple, rapid, and accurate, with good separation degree and high precision, which is suitable for the determination of residual solvents in sporoderm-broken *G. lucidum* spores powder [53].

6.3 Preclinical Toxicity of *Ganoderma lucidum* Spores Powder

To investigate the toxicological safety of *G. lucidum* spores powder, its toxicity was evaluated by taking acute toxicity test of mice, genetic toxicity test, and 30 days rat feeding trial [54–56].

6.3.1 Mouse Acute Toxicity

The mice were given gavage twice a day at an interval of 4 h, with a volume of 0.2 mL/10 g and a cumulative dose of 20.00 g/kg. It was observed for 2 consecutive weeks after intragastric administration, and the mice were weighed once a week. After the end of the experiment, the manifestations of poisoning and death were recorded, and anatomical observation was conducted. Subjects were evaluated according to acute toxicity grading criteria. After intragastric administration, the body weight was normal every week, no obvious poisoning symptoms were observed, and no death was observed on the 14th day. At the end of observation, the animals were sacrificed for anatomic examination, and the liver, spleen, kidney, stomach, intestine, heart, lung, and other main organs showed no obvious abnormal changes. The maximum tolerated dose (MTD) of mice was 20.0 g/kg, which indicated that samples were nontoxic product [54].

6.3.2 *Mouse Genotoxicity*

The results of salmonella reversion test, micronucleus test of bone marrow cells in mice (*G. lucidum* spores 2.5, 5, 10 g/kg) and mouse sperm malformation test (*G. lucidum* spores 2.5, 5, 10 g/kg) were all showed negative in genetic toxicity test [54].

6.3.3 *Rat Feeding Trial for 30 Days*

Eighty SD rats (each half male and female) were randomly divided into the control group and three *G. lucidum* spores dose groups (1.67, 3.33, 6.67 g/kg), which were equivalent to 25, 50, and 100 times of the recommended human dose, respectively. The *G. lucidum* spores were administered by adding the sample into the feed for 30 days. The control group was fed with basic feed. During the experiment, each rat was kept in a single cage and fed with water every day, with free activities, and food intake and food utilization were recorded. Blood samples were collected after feeding for 30 days, and the determination of hemoglobin, red blood cell volume, and counting of red blood cells, white blood cells, and platelets were carried out with fully automatic blood corpuscle counter. The serum biochemical parameters were determined by automatic biochemical analyzer. After the animals were sacrificed, gross anatomy was observed, and pathological sections of liver, kidney, stomach, spleen, ovary, testis, and duodenum were examined. During 30 days of feeding, no abnormal growth was observed in each group. Compared with the control group, there were no significant differences in body weight, food utilization rate, blood routine parameters, blood biochemical parameters, organ weight, etc., and no abnormal changes related to GLS feeding were found in gross anatomy and histopathological examination [54].

6.4 Pharmacological Effects of *Ganoderma lucidum* Spores Powder

6.4.1 Immunomodulatory Effect

6.4.1.1 Enhance the Immune Function of Normal Mice

The human recommended daily dose of *G. lucidum* spores powder is 1~4 g per day. Since the equivalent dose of mice is 10 times that of human body, 10 times of the human recommended daily dose is used as the test dose. The results showed that the *G. lucidum* spores powder significantly enhanced the delayed hypersensitivity of foot pad induced by sheep erythrocytes and the proliferation of splenic lymphocytes induced by ConA and increased the level of serum hemolysin antibody and the antibody production level of spleen cells, as well as promoted the carbon clearance and phagocytosis function in normal mice. The results suggest that human equivalent dose of *Ganoderma lucidum* spores powder can enhance cellular immunity, humoral immunity, and nonspecific immunity of normal mice [57–59].

6.4.1.2 Enhance the Immune Function of the Immunosuppression Mice

G. lucidum spores powder (1.2, 0.6, 0.3 g/kg by gavage once a day, for 15 days) was able to significantly increase the thymus index and spleen index and hemolysin level and enhance the phagocytosis of peritoneal macrophages and neutrophils of peripheral blood in the prednisone-immunosuppressive mice. Under physiological conditions, Th1/Th2 lymphocytes maintain a dynamic balance and maintain the normal function of cellular immunity and humoral immunity. *G. lucidum* spores powder could promote the Th1 cells secreting of IL-2 and INF-γ and Th2 cells producing IL-4. CD4+ cells have the function of assisting and inducing their precursor cells and play an active role in antitumor. CD8+ cells have negative regulatory effect and can inhibit B cells from producing antibodies. The constant CD4+/CD8+ ratio maintains the balance of cellular immune response. When the ratio of CD4+ and CD8+ increases, it indicates that the body has strong immune function. CD19+ is the surface marker of B cells, reflecting the level of humoral immunity. Compared with the prednisone-immunosuppressive group, the ratio of NK cells (CD49b+ cells), B cells (CD19+cells), and T cell subsets (CD3+, CD4+, and CD8+ cells) in peripheral blood was increased, and in particular, the ratio of CD4+/CD8+ was increased ($P < 0.05$) in each dose group of ganoderma spores. It is suggested that *G. lucidum* spores can positively regulate the immunosuppression induced by glucocorticoid in mice and can restore the specific and nonspecific immune functions inhibited by glucocorticoid [60].

Zhang W et al. (2018) observed the immunomodulatory effects of the wall-broken and wall-unbroken *G. lucidum* spores powder (0.75 g/kg/d by gavage for

30 days) on immunosuppressive mice induced by cyclophosphamide. Results showed that the wall-broken *G. lucidum* spores powder improved the survival rate and increased the femur nuclear cell number and the thymus index compared to the immunosuppressive model. Furthermore, the wall-broken *G. lucidum* spores powder extenuated the inhibition of proliferation of lymphocytes induced by cyclophosphamide and increased the serum levels of immunoglobulin and the number of lymphocytes in the immunocompromised mice. The results indicated that the wall-broken *G. lucidum* spores powder has obvious promoting effect on the immune function of immunocompromised mice [61].

6.4.2 Antitumor Effect

6.4.2.1 Inhibition of Tumor Cell Proliferation and Induction of Tumor Cell Apoptosis

In the past two decades, studies have shown that *G. lucidum* spores powder and its extracts, especially the triterpenoids, have an anti-proliferation and apoptosis-inducing effect on tumor cells and that systemic administration of *G. lucidum* spores powder has an inhibitory effect on tumor growth in xenograft of human tumor cells in immunodeficient nude mice. The mechanism may be related to the effect of triterpenes isolated from *G. lucidum* spores on the signal molecules involved in proliferation and apoptosis of tumor cells [62–64].

Chen JL et al. (2002) reported that using MTT and SRB method to study mixture of extract of *G. lucidum* and *G. lucidum* spores (MLGLGS) on growth inhibition of P388 murine leukemia cells and ten different types of tumor cell lines. It had found that the IC_{50} values of the mixture on P388, U-937, and HL-60 cells were all less than 2 mg/ml and that two human gastric cancer cell lines and two human lung cancer cell lines were all less than 4 mg/ml (Table 6.2). The growth inhibition rate of xenograft of human lung adenocarcinoma (1 g/kg) and gastric adenocarcinoma (2 g/kg) in nude mice was 49.47% and 43.09%, respectively [62].

Chen C et al. (2016) reported *G. lucidum* extracts and spore oil on tumor cells in vitro and in vivo. It was found that the IC_{50} values for *G. lucidum* extracts on HL60, K562, and SGC-7901 cell lines were 0.44 mg/ml, 0.39 mg/ml, and 0.90 mg/ml, respectively, and for *G. lucidum* spores oil 1.13 mg/ml, 2.27 mg/ml, and 6.29 mg/ml respectively. The inhibitory rates of *G. lucidum* extracts 1, 2, and 4 g/kg intragastric administration per day on S180-bearing mice were 15.6%, 19.0%, and 39.1%, respectively, and on H22-bearing mice were 28.8%, 38.5%, and 44.6%, respectively. The inhibitory rates of *G. lucidum* spores oil 1.2 g/kg intragastric administration per day on S180 and H22-bearing mice were 30.9% and 44.9%, respectively. *G. lucidum* extracts, but not *G. lucidum* spores oil, improved immunity indexes of tumor-bearing mice. *G. lucidum* extracts and spores oil inhibited activities of topoisomerase I and II in a dose-dependent manner, as seen by the increase of the ratio between supercoiled DNA and relaxed and broken DNA (Figs. 6.4 and 6.5).

Table 6.2 IC_{50} of MLGLGS on tumor cell lines Ref. [61]

Cell lines	IC_{50} (mg/ml)
Leukemia cell lines	
P388	0. 67
K562	6. 80
U937	1. 64
HL60	1. 44
Human gastroadenocarcinoma cell lines	
SGC7901	3. 50
MKN28	2. 81
Human colon cancer cell line	
HCT116	4. 10
Human hepatocarcinoma cell line	
BEL7402	6. 12
Human lung cancer cell lines	
SPC-A4	2. 31
A549	3. 27
Human ovarian cancer cell line	
HO-8910	3. 67

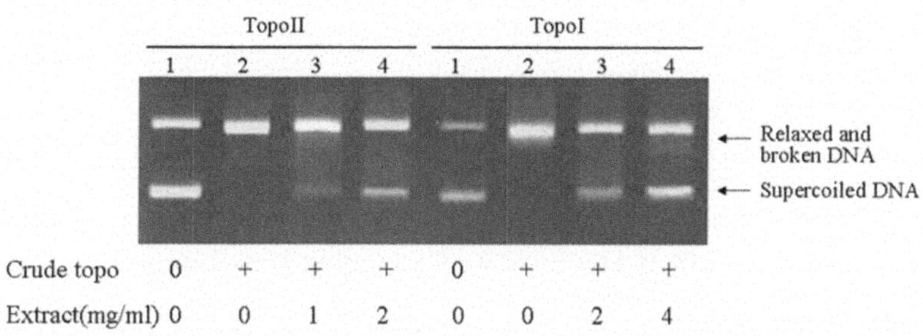

Fig. 6.4 Effects of *Ganoderma* extracts on Topo I- and II-mediated pBR322 DNA relaxation. Ganoderma extracts showed inhibitory effects on Topo I and II at a dose of 2 mg/ml. As dosage increased, 4 mg/ml extracts exhibited an even stronger effect on Topo I. Topo, topoisomerase [63]

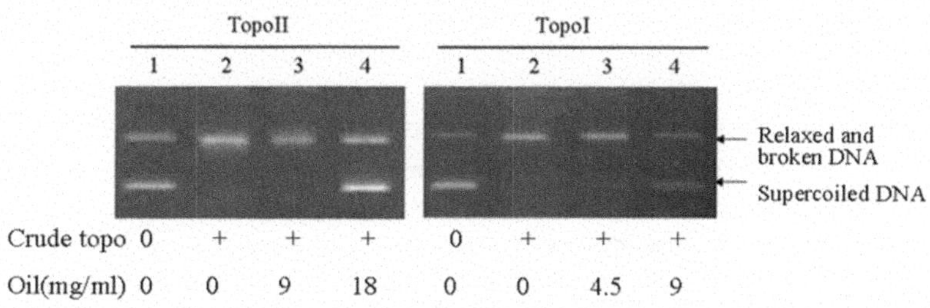

Fig. 6.5 Effects of Ganoderma spores oil on Topo I- and II-mediated pBR322 DNA relaxation. Ganoderma spores oil showed a dose-dependent inhibitory effect on Topo I. At the dose of 18 mg/ml, Topo II was also suppressed. Topo, topoisomerase [63]

G. lucidum spores oil was shown to block the cell cycle at the G1-to-S transition and induce a marked decrease of cyclin D1 level in K562 cells [63].

Li K et al. (2017) evaluated the in vitro antitumor activity of total triterpenes from sporoderm-broken spores of *G. lucidum* on human colon cancer HCT116, human lung cancer A549, and human breast cancer MDA-MB-231. The IC_{50} values of the total triterpenes on HCT116, A549, and MDA-MB-231 cell lines for 72 h exposure were 1.29, 1.68, and 3.81 mg/ml, respectively. HCT116 cells treated with the total triterpenes at 0.64~10 mg/ml for 48 h induced cell apoptosis remarkably [64].

Tan HS et al. (2018) recently studied the content of triterpenes in the fruiting bodies and spores of *G. lucidum* and compared the in vitro antitumor activities of triterpenoid-rich ethanol extracts from the fruiting bodies and spores of *G. lucidum*. The results showed that the triterpene content of *G. lucidum* fruiting bodies was 0.92% and that of *G. lucidum* broken spores was 2.95%. However, the lipid oil in *G. lucidum* broken spores greatly interfered with perchloric acid colorimetry. Therefore, the measured triterpene content of *G. lucidum* broken spores powder were probably much higher than the actual value. In the treatment with ethanol extract of fruiting bodies of *G. lucidum* at 100 μg/ml, the inhibition rates of human cervical cancer HeLa cells, breast cancer MCF-7 cells, and colon cancer HCT116 cells were 94.5%, 71.3%, and 20.7%, respectively. At the same concentration, the inhibition rates of the ethanol extracts of broken *G. lucidum* spores on HeLa, MCF-7, and HCT116 were 35.5%, 15.3%, and 12.6% respectively. The results indicated that the antitumor activity of ethanol extract from *Ganoderma lucidum* fruiting bodies was stronger than that of *G. lucidum* spores at the same concentration. The actual content of triterpenes in the two ethanol extracts needs to be further studied [65].

6.4.2.2 Antitumor Effect Mediated by Enhancing Host Immunity

In addition to inhibiting the tumor cell proliferation and/or inducing apoptosis in turn directly inhibiting or killing tumor cells, more importantly, *G. lucidum* spores can enhance host immune function to produce antitumor effect. Polysaccharides are one of the active components of *G. lucidum* in which antitumor effects have been extensively investigated and were mainly through immune-related mechanisms [66–69]. Recently several polysaccharides from broken spores of *G. lucidum* were isolated and have proved these polysaccharides have immunomodulatory activity and antitumor activity [70–72].

Sun LH et al. (2006) investigated the influence of *G. lucidum* spores on the dendritic cell (DC) derived from the bone marrow of H22 hepatoma-bearing mice, and its antitumor effect. The different dosages of *G. lucidum* spores (1, 2, 3 g/kg per day) were given to H22 hepatoma-bearing mice via gavage for 14 days continuously. After 24 h of the last taking dosage, the bone marrow was obtained with aseptic technique, and the GM-CSF and IL-4 were used to induce them into DC. Then the DC was cultured with different concentrations of *G. lucidum* spores (0.8, 3.2, 12.8 mg/L) in vitro. The results showed that *G. lucidum* spores could promote the proliferation and maturation of myelogenous DC and the expression of DC surface molecules such as CD11a and CD86 in tumor-bearing mice and enhance the cytotoxic T lymphocyte response induced by DC. Moreover, *G. lucidum* spores could increase the weight of spleen and thymus in tumor-bearing mice and significantly inhibit the growth of tumor. The results suggest that *G. lucidum* spores can enhance DC proliferation and promote its ability of antigen and improve the cellular immunity; thus it can inhibit the tumor growth [73].

Jiang Y et al. (2005) reported that the total polysaccharide was extracted from the spores of *G. lucidum*, and a polysaccharide named Lzps-1 with molecular weight of 8000 was obtained from the total polysaccharide. Lzps-1 was similar to Lentinan, all of which consisted of β-(1→3) glucose as the main chain and β-(1→6) glucose as a side-chain glucan, but Lzps-1 has an additional α-(1→4) glucose in the side chain of β-(1→6) glucose. The total polysaccharide (100, 200 mg/kg by intraperitoneal injection) inhibited the growth of Lewis lung cancer in mice by 62.89% and 53.03%, respectively, and that of sarcoma S180 in mice was 57.36% and 46.79%, respectively; moreover the activity of NK cells in Lewis lung cancer-bearing mice was enhanced [74].

Wang PY et al. (2012) found that broken spore polysaccharide (Gl-BSP) extracted from spores of *G. lucidum* (50, 100, and 200 mg/kg by intragastric administration) significantly inhibited S180 sarcoma growth in mice; however, Gl-BSP did not inhibit sarcoma180 and PG cell proliferation in vitro. The authors also found that the serum levels of IL-2, IFN-γ, and TNF-α were markedly increased in S180-bearing mice administrated with Gl-BSP (200 mg/kg), compared with that in S180-bearing control mice. The cytokine-neutralizing antibodies such as TNF-α or IFN-γ antibody apparently diminished S180 cell growth inhibition produced by the serum from mice treated with Gl-BSP 200 mg/kg, suggesting the cytokines in the serum of Gl-BSP-treated mice may play at least part role in antitumor effect of Gl-BSP. Further research also found that Gl-BSP promoted the splenic lymphocyte proliferation induced by Con A or LPS, enhanced nature killer cell (NK cell) cytotoxic activity, augmented the percentage of neutral red phagocytosis by macrophages, and increased the percentage of the $CD4^+$ and $CD8^+$ subset in S180-bearing mice. These findings suggest that the antitumor activity of Gl-BSP may be mainly related to the activation of the immune response of the host organism by stimulation of NK cells, T cells, and macrophages [75].

Su JY et al. (2018) reported that an extract derived from the sporoderm-breaking spores of *G. lucidum* (ESG) was able to suppress 4 T1 tumor growth in vivo rather than in vitro. Flow cytometry analysis revealed that ESG could significantly increase both Tc subpopulation and the ratio of Tc/Th in peripheral blood of the tumor-bearing mouse; similar promotion on Tc and the ratio of Tc/Th was also found in tumor-infiltrating lymphocyte. Moreover, ESG evidently down-regulated the two immune checkpoints, programmed cell death protein-1 (PD-1, in the spleen) and cytotoxic T lymphocyte antigen-4 (CTLA-4, in the tumor), suggesting that ESG could effectively restore the T cell paradigm by recovering the exhaustion status via suppressing the co-inhibitory checkpoints. Taken together, ESG would serve as a natural anticancer adjuvant via a restoration on the exhausted Tc, highlighting important clinical implications for the treatment of breast cancer [76].

6.4.2.3 Suppression on Tumor Metabolism and Reshaping Gut Microbiota

Su JY et al. (2018) found that the combination of paclitaxel (PTX) and spores of *G. lucidum* (SGP) displayed an improved tumor control, in which mRNA expression of several Warburg effect-related proteins, i.e., glucose transporter 3 (Glut3), lactate dehydrogenase A (Ldha), and pyruvate dehydrogenase kinase (Pdk), and the metabolite profile of tumor was evidently altered. Flow cytometry analysis revealed that the combination treatment recovered the exhausted tumor infiltration lymphocytes (TILs) via inhibiting the expressions of immune checkpoints (PD-1 and Tim-3), while PTX alone evidently increased that of CTLA-4. 16S rRNA sequencing revealed a restoration by the combination treatment on gut microbiota dysbiosis induced by PTX, especially that *Bacteroides*, *Ruminococcus*, and other five genera were significantly enriched, while the cancer-risk genera, *Desulfovibrio* and *Odoribacter*, were decreased. Moreover, spearman correlation analysis showed that abundance of *Ruminococcus* was significantly negative-associated with the amount of frucotose-6-phosphate within the regulation of tumor metabolism and gut microtumor. Collectively, the study suggests the clinical implication of SGP as an adjuvant candidate for PTX against breast cancer, which possibly relies on the regulation of tumor metabolism and gut microbiota [77].

6.4.3 *Neuroprotective Effect*

6.4.3.1 Inhibition of Apoptosis of Nerve Cells and Protection for Nerve Cells

Xie AM et al. (2004) used microinjection of 6-hydroxydopa (6-OHDA) into midbrain substantia nigra to induce Parkinson's disease (PD) animal model. It was found that *G. lucidum* spores powder could protect substantia nigra neurons and weaken the damage of neurons. Electron microscopic observation showed that there were more apoptotic neurons in PD group with unclear mitochondrial cristae and dilated endoplasmic reticulum (ER), while in *G. lucidum* spores powder pretreatment group, there were only slight morphological changes of mitochondria, and apoptotic neurons were occasionally observed. The results showed that *G. lucidum* spores powder could inhibit the apoptosis of nerve cells and might have protective effect on the brain [78].

6.4.3.2 Protection of Damaged Spinal Motoneurons

Zhang W et al. (2006) found that *G. lucidum* spores powder (GASP) 8 g/kg could promote the survival and axon regeneration of injured spinal motoneurons in rats. There were six kinds of proteins differentially expressed between the treatment group and control group, which were collapsin response mediator protein 2 (CRMP-2), F-actin capping protein beta subunit (FCP-β), isocitrate dehydrogenase [NAD] subunit beta (IDH-β), ATPase, glutamate oxaloacetate transaminase-1 (GOT1), and M2 pyruvate kinase (M2-PK). The expression levels of CRMP-2, IDH-β, ATPase, and GOT1 were higher in the GASP-treated group than those in the untreated group, while the expression levels of FOP-β and M2-PK were lower than those in the untreated group. *G. lucidum* spores treatment group could promote the expression of CRMP-2 in spinal cord collapse response, and CRMP-2 could promote the growth of axons by regulating the formation of microtubule of neurons. However, the expression of FCP-β decreased significantly in GASP-treated group, and FCP-β could not prolong the microtubule by "capping" the growing microtubule. *G. lucidum* spores also promoted the expression of isocitrate dehydrogenase β (IDH-β), adenosine triphosphatase (ATPase), and glutamic oxaloacetate aminotransferase 1 (GOT1) in injured spinal cord and inhibited pyruvate kinase M2 (M2-PK) expression. Therefore, *G. lucidum* spores may promote axon regeneration of injured spinal motoneurons by regulating the expression of CRMP-2/FCP-β. By increasing the expression of IDH-β, ATPase, and GOT1, the tricarboxylic acid cycle of mitochondria of injured spinal motor neurons was promoted, and the synthesis of ATP was increased. By reducing the expression of M2-PK and reducing the glycolysis of glucose, glycogen can be efficiently utilized to promote the survival of neurons and the regeneration of axons. The results indicated *G. lucidum* spores may promote the survival and axon regeneration of injured spinal motor neurons in rats by virtue of up- or down-regulating the expression levels of the proteins mentioned above [79].

6.4.3.3 Anti-Alzheimer's Disease (AD)

Alzheimer's disease is the most common neurodegenerative disease in the elderly. Its clinical manifestations are memory impairment and other cognitive impairment. The pathological changes could be seen in senile plaques (SP) and neuronal fiber tangles (NFTs) as well as in the decrease of the number of neurons and granular vacuolar degeneration (GVD). It has been proved that the formation of SP and NFTs in the brain of AD is related to the free radicals. Oxidative damage is related to oxidative stress, so oxidative stress and free radical damage are important mechanisms for the pathogenesis of AD. At present, the treatment of AD is limited to improving the patient's symptoms and delaying the progress of the disease. A series of pharmacological studies have found that *G. lucidum* spores has anti-AD effect.

Zhou Y et al. (2012) reported that after intraventricular injection (ICV) of streptozotocin (STZ), the content of methane dicarboxylic aldehyde (MDA) in the hippocampus of the model rats was significantly increased, and the glutathione reductase (GR), reduced GSH, adenosine triphosphate (ATP), and cytochrome oxidase (CytOx) was decreased in the hippocampus of the model rats. There was a significant decrease in learning and memory ability and severe damage to hippocampal neurons in the model rats. Oral administration of *G. lucidum* spores powder 2, 4, and 8 g/kg twice a day for 21 days before modeling could significantly reverse the cognitive behavior, oxidative stress index, energy metabolism disorder, and histopathological changes of hippocampus induced by ICV STZ in rats. The results showed that *G. lucidum* spores powder could prevent oxidative stress damage and energy metabolism disorder in ICV STZ rats and provide pharmacological basis for the prevention and treatment of AD with *G. lucidum* spores powder [80].

Recently, Qin C et al. (2017) observed the effect of *G. lucidum* spores powder and spore oil on pathological changes in APP/PS-1 transgenic AD mice. 4 months APP/PS-1 transgenic mice were divided into AD model group, *G. lucidum* spores powder and spores oil treatment groups, and positive control group with donepezil hydrochloride. C57BL/6J wild mice were used as normal group. After 4-month treatment, the mice were sacrificed, and histopathological morphological and structural differences in each group were observed. The results showed that senile plaques (SP) and NFTs were degraded, decreased, or disappeared in the rats treated with *G. lucidum* spores powder and spores oil, and amyloid angiopathy also significantly reduced. The number of immature neurons in hippocampus and dentate nucleus in treatment groups was significantly higher than that in AD model group. The DCX immunohistochemical staining in treatment groups was positive, indicating that cerebellar neural stem cells appeared early and in large quantity. The hippocampal nerve cells in the group treated with *G. lucidum* spores powder and spores oil were observed by electron microscopy to be more complete than that in the AD model mice, with intact nuclear membrane, complete mitochondria, endoplasmic reticulum, Golgi body, microtubule structure, and synapses in the cytoplasm. Microglia were normal. The cell and histological toxicity of *G. lucidum* spores powder and spores oil were not observed in the treatment groups. The results indicated that *G. lucidum* spores powder and spores oil could reduce SP, NFTs, and amyloid angiopathy and have the protection of neuronal cell and promote the growth of immature nerve cells in the brain of AD rats [81].

6.4.3.4 Sedation and Improving Sleep

Zhang ML et al. (2013) found that sporoderm-breaking spores of *G. lucidum* 0.25, 0.5, and 1.0 g/kg intragastric administration once a day for 30 days could significantly prolong the sleep time induced by pentobarbital sodium, shorten the sleep latency induced by barbitone sodium, and increase the incidence of sleep induced by pentobarbital sodium. However, there was no significant effect of spores of *G. lucidum* on the number of sleeping animals and sleep time in direct sleep mice [82].

6.4.3.5 Antiepileptic Action

Epilepsy is a group of chronic brain diseases caused by abnormal firing of highly synchronized neurons in the brain caused by different etiologies. It is characterized by central nervous system dysfunction caused by transient and frequent epileptic seizures. Nitric oxide (NO), nitric oxide synthase (NOS), IL-2, and Ca2+ are involved in the kindling of epilepsy. Apoptosis after epilepsy is one of the important links leading to brain injury. Therefore, inhibition of apoptosis is of great significance for the prevention and treatment of brain tissue damage caused by epilepsy. In recent years, studies have shown that administration of *G. lucidum* spores powder at 150~300 mg/kg by gavage can significantly reduce the contents of IL-2, NO, NOS, and intracellular Ca^{2+}, lower the levels of caspase-3 and caspase-9, and inhibit the apoptosis of neurons in epileptic rats, which play an anti-epileptic role [83–85].

6.4.4 Improve Liver Function

Sporoderm-breaking spores of *G. lucidum* 0.50 g/kg and 3.0 g/kg administered by gavage significantly reduced collagen deposition of liver in CCl_4-induced liver fibrosis model rats, decreased the expression of TGF-1 and TIMP-1, inhibited collagen synthesis, and promoted collagen decomposition, thereby inhibited the formation of liver fibrosis [86]. *G. lucidum* spores oil 1.00 g/kg, 0.33 g/kg, and 0.17 g/kg administered by gavage for 2 weeks significantly reduced aspartic acid transaminase (AST) and alanine aminotransferase (ALT) activity in CCl_4-induced liver injury model mice [87]. *G. lucidum* spores powder 0.5 and 1.0 g/kg administered by gavage for 2 weeks could significantly inhibit CCl_4-induced liver fibrosis in mice. Compared with the model group, serum ALT, AST levels, and MDA levels in liver tissues were significantly decreased, and MMp-9 protein expression in liver tissues was significantly inhibited in the *G. lucidum* spores powder group [88]. *G. lucidum* spores powder 1.0 g/kg administered by gavage for 10 days in mice of sub-acute alcoholic hepatic injury significantly decreased levels of serum ALT and AST, and the hepatic pathological changes were attenuated or even reversed [89]. Con A 5 mg/kg tail intravenous injection was given once every other day for 2 weeks to establish a model of immune liver injury in male Balb/c mice. Broken *G. lucidum* spores powder 1.0 g/kg and 2.0 g/kg were given by gavage for 4 weeks in mice with immune liver injury. Compared with the model group, the serum ALT and AST levels of the mice in the broken *G. lucidum* spores powder group were significantly decreased, and the degree of liver cell degeneration, necrosis, and inflammatory cell infiltration were significantly reduced. Therefore, broken *G. lucidum* spores can improve liver function and inflammation in Con A-induced immune liver injury mice [90].

6.4.5 Antioxygenation

The effect of the water-soluble preparation of the spores of *G. lucidum* on spontaneous and Fe^{2+}-induced lipid peroxidation of skeleton muscle homogenate was studied by Gu X et al. (1993). The results showed that addition of the preparation of spores to the incubation mixture inhibited the production of MDA in both spontaneous and Fe^{2+}-induced lipid peroxidation of skeleton muscle homogenate from mice in vitro. Intraperitoneal injection of the preparation of spores significantly inhibited the increase of MDA in serum induced by subcutaneous injection of 2,4-dichlorophenoxyacetic acid in mice and increased the catalase activity in liver cytosol in normal mice. In addition, the preparation of spores could directly inhibit the production of superoxide anion in xanthine oxidase/xanthine system. These results indicated that *G. lucidum* spores powder could inhibit the lipid peroxidation of muscle cell membrane caused by reactive oxygen species and protect muscle cells from injury. This may be related to its therapeutic effect on multiple myositis and other diseases [91].

Zhang SQ et al. (2003) observed protective effect of Jisheng injection (an extract of *G. lucidum* spores) on Langendorff isolated rat's heart and antioxidation effects in vitro. Forty minutes after perfusion, the cardiac function, coronary flow, and myocardial water content were determined, and lactate dehydrogenase (LDH), creatine kinase (CK) activity in perfusate, SOD activity, MDA content in myocardial tissue, and pathologic change in myocardium were also observed. The results showed that the cardiac function and coronary flow of isolated heart preserved in modified Euro-Collins solution containing Jisheng injection was significantly better than those in the control ($P < 0.01$), with the LDH, CK activity, and MDA content significantly lower ($P < 0.01$ and $P < 0.05$), SOD activity significantly higher ($P < 0.05$), and pathologic injury milder than those in control, but comparison of cardiac water content between the two groups showed insignificant difference. It was suggested that the protective effect of *G. lucidum* spores powder extract on the heart may be related to its antioxidant effect and the prevention of oxygen free radical damage [92].

Oxidative stress is one of the pathogenesis of diabetic nephropathy. The generation of oxygen free radicals in hyperglycemia and the lipid peroxidation reaction caused by oxygen free radicals play an important role in the kidney damage of diabetic patients. Liu Y et al. (2008) explored the protective effect of *G. lucidum* spores on oxygen free radical injury in streptozotocin-induced diabetic rats. The results showed that compared with model group, *G. lucidum* spores (250 mg/kg once a day for 10 weeks) feeding group significantly increased SDH, SOD, and GSH-PX activity and reduced MDA content in kidney tissue. It was suggested that *G. lucidum* spores could improve the level of antioxidant enzyme and reduce the content of lipid peroxide in renal tissue, protect the damage of free radicals to the kidney, and play a protective role in the kidney of type 2 diabetic rats [93].

Hu ZM et al. (2016) investigated the protective effect of *G. lucidum* spores on stomach inflammation induced by ethanol in mice. The results showed that *G. lucidum* spores (0.5 g/kg and 1 g/kg) could alleviate the pathological changes of gastric ulcer induced by ethanol; effectively decrease the gastric ulcer index in mice; significantly increase the level of SOD and decrease the level of MDA, IL-6, and TNF-α in serum; and reduce the protein expression of NF-κB P65, COX-2. At the same time, *G. lucidum* spores also significantly increased SOD activity in stomach tissue. These findings suggest that *G. lucidum* spores had a protective effect on ethanol-induced acute gastric ulcer, maybe mainly through antioxidant and anti-inflammatory mechanisms [94].

6.4.6 *Anti-radiation and Anti-mutation Effect*

In order to assess the radioprotective effect of *G. lucidum*, mice were given different doses (0, 0.25, 0.50, 1.50 g/kg) of *G. lucidum* spores powder by gavage for 60 days. On the 30th day, each animal was irradiated once with 8.2 Gy 60 Co γ-ray. By the 60th day, the survival rates, the average survival time and the average white blood cell counts were observed. It was found that the three parameters were all much higher in the groups treated with *G. lucidum* spores powder than those in the control group, suggesting significant protective effect of *G. lucidum* spores against radiation [95–97].

In addition, the micronuclei frequency induced by cyclophosphamide in mouse bone marrow cells and chromosome aberration frequency induced by mitomycin C were both inhibited significantly by *G. lucidum* spores powder [98, 99].

6.5 Clinical Application of *Ganoderma Lucidum* Spores Powder

Previous studies have suggested that Ganoderma products are considerable use in the adjuvant treatment of diverse cancers in combination with surgery, chemotherapy, and radiotherapy. The detailed efficacies are as follows: (1) stimulating the immune system improving the therapeutic effect of tumor, (2) reducing the side effects and elevating the tolerance of cancer patients receiving chemo- or radiotherapy, and (3) improving the survival quality of cancer patients.

6.5.1 *Stimulating the Immune Function and Improving the Therapeutic Efficacy of Chemotherapy*

Qi YF et al. (1999) reported the adjuvant chemotherapy by *G. lucidum* spores for the treatment of 200 cases of digestive tract cancer patients. Test group (100 cases) was oral administration of *G. lucidum* spores at a dosage of 1 g three times daily for 8 weeks. Control group (100 cases) was oral Zhenqi Fuzheng granule at a dosage of 15 g three times daily for 8 weeks. Conventional chemotherapy should be given in both groups at the beginning of each course of treatment. The results indicated that adjuvant chemotherapy by *G. lucidum* spores in digestive tract cancer patients could enhance the immune function of patients, improve the total effective rate of chemotherapy, reduce the bone marrow suppression of chemotherapy drugs, and improve the patient's body weight and Karnofsky score [100].

Zhen ZJ et al. (2012) evaluated the clinical efficacy of *G. lucidum* spores (GLS) on recurrence after curative resection of hepatocellular carcinoma (HCC). Sixty patients undergoing curative resection of HCC were randomly divided into conventional treatment group and GLS treatment group. After 2 years of follow-up, the disease-free survival (DFS) and overall survival (OS) were compared between the two groups.Results showed that the 2-year DFS were 53.3% and 70.0%($P = 0.034$), and 2-year OS were 60.0% and 83.3% ($P = 0.023$), respectively, in the conventional treatment group and GLS group. There were no significant differences in postoperative complication rate between the two groups ($P = 0.472$), and only one case developed diarrhea due to adverse effect of GLS. This finding indicates that GLS treatment is safe and effective to inhibit tumor recurrence and improve DFS and OS of the patients after radical resection for HCC [101].

Furthermore, Zhen ZJ et al. (2013) also probed into whether GLS can improve the immunological function of patients with hepatocellular carcinoma (HCC) after operation. The results showed that compared with the control group, the percentage of CD4+ cells and percentage of NK cells in HCC patients significantly reduced, and percentage of CD8+ cells in HCC patients was significantly elevated. Before the operation, there was no significant difference in the percentages of CD4+, CD8+, and NK cells of patients between the conventional therapy control group and the GLS therapy group. The percentages of CD4+, CD8+, and NK cells of patients in two groups evidently decreased 1 day after operation compared with the preoperative data ($P < 0.01$). Compared with the conventional therapy group, the percentage of CD4+ in the GLS therapy group was significantly elevated 7 and 28 days after operation ($P < 0.05$, while the percentage of CD8+ was significantly decreased ($P < 0.05$), and the NK cells was significantly elevated ($P < 0.05$) (Table 6.3). It suggests that cellular immunological function in patients with HCC is inhibited before and after the operation. Application of *G. lucidum* spores early after the operation can improve the cellular immunological function and help to maintain the body immunological balance [102].

6.5.2 Reducing the Side Effects and Enhancing the Tolerance of Cancer Patients to Chemotherapy

Wang J et al. (2016) observed the clinical efficacy of *G. lucidum* spores capsule combined with chemotherapy in the treatment of non-small cell lung cancer (NSCLC) and its effect on patients' immune function. NSCLC patients (134 cases) were randomly divided into the adjuvant treatment group and the chemotherapy control group. The chemotherapy control group was treated with conventional chemotherapy, and the adjuvant treatment group was treated with *G. lucidum* spores capsules combined with conventional chemotherapy. The clinical efficacy and immune function of the two groups were observed. The results showed that the effective rate and total effective rate in the adjuvant treatment group were significantly higher than those in the chemotherapy control group ($P < 0.05$), while the incidence of myelosuppression and gastrointestinal reaction was significantly lower than that of the chemotherapy control group ($P < 0.05$). In the adjuvant treatment

Table 6.3 Peripheral blood T lymphocyte subsets and NK cell percentage in two groups of liver cancer patients (%) Ref. [102]

		CD4+ cells				CD8+ cells				NK cells			
Groups	n	d0	d1	d7	d28	d0	d1	d7	d28	d0	d1	d7	d28
Control	**35**	**34 ± 7**	**29 ± 4**	**33 ± 5**	**38 ± 6**	**30 ± 4**	**28 ± 4**	**33 ± 5**	**29 ± 3**	**14 ± 4**	**12 ± 4**	**15 ± 3**	**16 ± 4**
GLS	**35**	**34 ± 7**	**30 ± 3**	**37 ± 4**	**42 ± 7**	**31 ± 3**	**28 ± 3**	**29 ± 3**	**27 ± 3**	**13 ± 4**	**10 ± 3**	**17 ± 3**	**18 ± 4**
t value		0.00	1.18	3.70	2.39	1.18	0.00	4.06	2.79	1.05	1.27	2.79	2.09
p value		1.00	>0.05	<0.05	<0.05	>0.05	1.00	<0.05	<0.05	>0.05	>0.05	<0.05	<0.05

Note: d0: before the operation; d1, d7, d28: day 1, day 7, day 28 after operation

group, CD3+, CD4+, CD8+, and CD4+/CD8+ were significantly increased ($P < 0.05$) and significantly better than the chemotherapy group ($P < 0.05$). This clinical observation suggests that chemotherapy combined with compound *G. lucidum* spores capsule in the treatment of NSCLC patients could effectively improve the clinical efficacy, reduce the adverse reactions caused by chemotherapy, and significantly improve the immune function of patients [103].

6.5.3 Improving the Survival Quality of Cancer Patients

Zhao H et al. (2012) investigated the effectiveness of *G. lucidum* spores powder on cancer-related fatigue in breast cancer patients undergoing endocrine therapy. Breast cancer patients (48 cases) with cancer-related fatigue symptoms who were treated with endocrine therapy were randomly divided into a trial group and a control group. Patients in the trial group received oral *G. lucidum* spores powder for 1 g each time, 3 times a day for a total of 4 weeks. The control group was treated with a placebo. Before and after treatment, patients were assessed on Functional Assessment of Cancer Therapy-Fatigue (FACT-F), hospital anxiety and depression scale (HADS), and EORTC quality of life questionnaire (EORTC qlq-c30). TNF-α and IL-6 levels in blood and liver and kidney function were measured before and after taking *G. lucidum* spores powder. FACT-F evaluation results showed that compared with the control group, the body condition score and fatigue degree score of the patients receiving *G. lucidum* spores powder were significantly increased ($P < 0.01$ and $P < 0.01$, respectively). HADS and EORTC qlq-c30 results showed that the degree of anxiety and depression in the trial group was reduced and the life quality satisfaction was increased ($P < 0.01$ and $P < 0.01$, respectively). TNF-α and IL-6 in the blood of the patients in the trial group were significantly decreased after treatment ($P < 0.01$, $P < 0.05$). The results showed that *G. lucidum* spores powder could improve the cancer-related fatigue, depression, and quality of life of breast cancer patients receiving endocrine therapy. The therapeutic mechanism may be related to the reduction of TNF-α and IL-6, and cancer-related fatigue markers, in breast cancer patients undergoing endocrine therapy. No serious adverse reactions occurred in the course of taking *G. lucidum* spores powder [104].

6.6 Conclusion

The wall breaking technology of spores promoted the release of effective components from *Ganoderma lucidum* spores, and the released content of triterpenes and polysaccharides in the wall breaking *G. lucidum* spores was significantly higher than that of unbroken spores. The methods for breaking wall of *G. lucidum* spores include physical, physicochemical, enzymatic, and zymotic. Among them, a physical method of high-speed centrifugal shearing of pulverizer can make the *G. lucidum* spores wall broken rate reached almost 100% [7].

As for the study of the chemical composition of *G. lucidum* spores, a lot of research has focused on polysaccharides and triterpenes. Due to the heterogeneity of polysaccharide structure and molecular weight and the difficulty in structural identification, there are a few reports on the pure polysaccharides obtained from *G. lucidum* spores in which most polysaccharides are glucan composed of different glucose combinations [22–28]. In the study of polysaccharide activity of *G. lucidum* spores, it is mainly to study the total polysaccharide of *G. lucidum* spores or polysaccharide components within a certain molecular weight range. At present, 29 triterpenes have been isolated from *G. lucidum* spores, the amount of which is far less than that of triterpenes obtained from the fruiting bodies of *G. lucidum*. Ganosporelactone A, ganosporelactone B, and lucidenic acid SP1 have been isolated and identified only in spores of *G. lucidum*, but not in the fruiting bodies of *G. lucidum* [15–21]. In addition, about 18 steroidal compounds [18, 31–35] and 15 fatty acids were obtained from Ganoderma spores. Due to the small amount of single compounds obtained (most single compounds do not have detailed biological activity data), only a few single compounds has a limited in vitro biological activity data. There are still a lot of studies in animal models in vivo need to do.

Although preclinical studies have shown that *G. lucidum* spores have many pharmacological effects such as enhancing host immune function and antitumor activity, reducing side effects of radiotherapy and chemotherapy, scavenging free radicals, antioxidation, protective cell function, sedation and improvement of sleep, anti-Alzheimer's disease, and anti-epilepsy. However, existing clinical studies mainly focus on adjuvant chemo- or radiotherapy, enhancing the immune function of patients, producing synergy effect with chemo- or radiotherapy, reducing the toxic and side effects of chemo- or radiotherapy, and improving the quality of life of patients. There are still many pharmacological effects of *G. lucidum* spores that need to be verified by clinical studies.

References

1. Zhao D, Chang MW, Li JS, Suen W, Huang J (2014) Investigation of ice-assisted sonication on the microstructure and chemical quality of *Ganoderma Lucidum* spores. J Food Sci 79(11):2253–2265
2. Liu X, Wang JH, Yuan JP (2005) Pharmacological and anti-tumor activities of Ganoderma spores processed by top-down approaches. J Nanosci Nanotechnol 5(12):2001–2013
3. Chen T, Li K, Chen C, Chen X, Wen Q, Lin X (1998) Ultrastructure of basdiospores and sporoderm of *Ganoderma Lucidum* and *G. sinense*. Fujian J Agric Sci 4:33–38
4. National standard of the People's Republic of China (GB/T 29344–2012), Standards on harvesting and processing techniques of *Ganoderma Lucidum* spore powder [S]
5. Liu X, Yuan JP, Chung CK, Chen XJ (2002) Antitumor activity of the sporoderm-broken germinating spores of *Ganoderma Lucidum*. Cancer Lett 182(2):155–161

6. Xie YZ, Li SZ, Yee A, La Pierre DP, Deng Z, Lee DY, Wu QP, Chen Q, Li C, Zhang Z, Guo J, Jiang Z, Yang BB (2005) *Ganoderma Lucidum* inhibits tumour cell proliferation and induces tumour cell death. Enzym Microb Technol 40(1):177–185
7. Yue GGL, Fung KP, Leung PC, Lau CBS (2008) Comparative studies on the immunomodulatory and antitumor activities of the different parts of fruiting body of *Ganoderma Lucidum* and Ganoderma spores. Phytother Res 22(10):1282–1291
8. Ma J, Fu Z, Ma P, Su Y, Zhang Q (2007) Breaking and characteristics of *Ganoderma Lucidum* spores by high speed centrifugal shearing pulverizer. J Wuhan Univ Technol 22(4):617–621
9. Wang JJ, Ke R, Zhang SY (2017) Breaking the sporoderm of *Ganoderma Lucidum* spores by combining chemical reaction with physical actuation. Nat Prod Res 31(20):2428–2434
10. Fu YJ, Liu W, Zu YG, Shi XG, Liu ZG, Schwarz G, Efferth T (2009) Breaking the spores of the fungus *Ganoderma Lucidum* by supercritical CO_2. Food Chem 112(1):71–76
11. Guo L, Xie J, Ruan Y, Zhou L, Zhu H, Yun X, Jiang Y, Lu L, Chen K, Min Z, Wen Y, Gu J (2009) Characterization and immunostimulatory activity of a polysaccharide from the spores of *Ganoderma Lucidum*. Int Immunopharmcol 9(10):1175–1182
12. Wu QP, Xie YZ, Li SZ, La Pierre DP, Deng ZQ, Chen Q, Li C, Zhang Z, Guo J, Wong CKA, Lee DY, Yee A, Yang BB (2006) Tumour cell adhesion and integrin expression affected by *Ganoderma Lucidum*. Enzym Microb Technol 40(1):32–41
13. Chaiyasut C, Kruatama C, Sirilun S (2010) Breaking the spores of *Ganoderma Lucidum* by fermentation with lactobacillus plantarum. Afr J Biotechnol 9(43):7379–7382
14. Ma J, Fu Z, Ma P, Su Y, Zhang Q (2007) Breaking and characteristics of *Ganoderma Lucidum* spores by high speed entrifugal shearing pulverizer. J Wuhan Univ Technol 22(4):617–621
15. Chen RY, Yu DQ (1991) Studies on the triterpenoid constituents of the spores from *Ganoderma Lucidum* karst. Acta Pharm Sin 26(4):267–273
16. Chen RY, Yu DQ (1991) Application of 2D NMR techniques in the structure determination of Ganosporelactone A and B. Acta Pharm Sin 26(6):430–436
17. Zhang XQ, Pan GL, Cheng Y, Wang Y, Ye WC (2008) Chemical constituents of the spores of *Ganoderma Lucidum*. J Chin Med Mat 31(1):41–44
18. Zhang CR, Yang SP, Yue JM (2008) Sterols and triterpenoids from the spores of *Ganoderma Lucidum*. Nat Prod Res 22(13):1137–1142
19. Min BS, Gao JJ, Nakamura N, Hattori M (2000) Triterpenes from the spores of *Ganoderma Lucidum* and their cytotoxicity against meth-A and LLC tumor cells. Chem Pharm Bull 48(7):1026–1033
20. Min BS, Nakamura N, Miyashiro H, Bae KW, Hattori M (1998) Triterpenes from the spores of *Ganoderma Lucidum* and their inhibitory activity against HIV-1 protease. Chem Pharm Bull 46(10):1607–1612
21. Min BS, Gao JJ, Hattori M, Lee HK, Kim YH (2001) Anticomplement activity of terpenoids from the spores of *Ganoderma Lucidum*. Planta Med 67(9):811–814
22. Bao XF, Duan JY, Fang XY, Fang JI (2001) Chemical modifications of the (1→3)-α-D-glucan from spores of *Ganoderma Lucidum* and investigation of their physicochemical properties and immunological activity. Carbohydr Res 336:127–140
23. Bao XF, Fang JN (2001) Isolation and structural determination of a glucan from the spores of *Ganoderma Lucidum*. Acta Bot Sin 43(3):312–315
24. Bao XF, Liu CP, Fang JN, Li XY (2001) Structural and immunological studies of a major polysaccharide from spores of *Ganoderma Lucidum* (Fr.) Karst. Carbohydr Res 332(1):67–74
25. Dong Q, Wang Y, Shi L, Yao J, Li J, Ma FL, Ding K (2012) A novel water-soluble β-D-glucan isolated from the spores of *Ganoderma Lucidum*. Carbohydr Res 353:100–105
26. Guo L, Xie JH, Ruan YY, Zhou L, Zhu HY, Yun XJ, Jiang Y, Lu L, Chen KL, Min ZH, Wen YM, Gu JX (2009) Characterization and immunostimulatory activity of a polysaccharide from the spores of *Ganoderma Lucidum*. Int Immunopharmacol 9(10):1175–1182
27. Zhang LX, Zhang YJ, Zhang LP (2011) Structure and immunological activity of a novel polysaccharide from the spores of *Ganoderma Lucidum*. Afr J Biotechnol 10(53):10923–10929

28. Wang YT, Liu YF, Yu HZ, Zhou S, Zhang Z, Wu D, Yan MQ, Tang QJ, Zhang JS (2017) Structural characterization and immuno-enhancing activity of a highly branched water-soluble β-glucan from the spores of *Ganoderma Lucidum*. Carbohydr Polym 167:337–344
29. Chen TQ, Li KB, Xu J, Zhu PG, Zheng Y (1997) Amino acids, fatty acids and element composition of the spores of log-cultivated *Ganoderma Lucidum*. Mycosystema 16(1):52–56
30. Zhang NR, Zhang XY (1997) Analysis of vitamins and polysaccharide in the spore powder of *Ganoderma Lucidum* (Leyss. Ex Fr.) Karst. Chin J Biochem Pharm 18(1):37–38
31. Chen RY, Wang YH, Yu DQ (1991) Studies on the chemical constituents of the spores from *Ganoderma Lucidum*. Acta Bot Sin 33(1):65–68
32. Zhang XQ, Pang GL, Cheng Y, Wang Y, Ye WC (2008) Chemical constituents of the spores of *Ganoderma Lucidum*. J Chin Med Mater 31(1):41–44
33. Weng YF, Xiang L, Matsuura A, Zhang Y, Huang QM, Qi JH (2010) Ganodermasides A and B, two novel anti-aging ergosterols from spores of a medicinal mushroom *Ganoderma Lucidum* on yeast via *UTH1* gene. Bioorg Med Chem 18:999–1002
34. Weng YF, Lu J, Xiang L, Matsuura A, Zhang Y, Huang QM, Qi JH (2011) Ganodermasides C and D, two new anti-aging ergosterols from spores of the medicinal mushroom *Ganoderma Lucidum*. Biosci Biotechnol Biochem 75(4):800–803
35. Ge FH, Duan MH, Li J, Shi QL (2017) Ganoderin A, a novel 9,11-secosterol from *Ganoderma Lucidum* spores oil. J Asian Nat Prod Res 19(12):1252–1257
36. Hou CY, Sun YT, Yang L, Xue H (1988) Studies on the chemical constituents of the spores from *Ganoderma Lucidum* (Leyss. Et Fr.) Karst. Acta Bot Sin 30(1):66–70
37. Chen TQ, Wu JG, Wu YB, Wang H, Mao FH, Wu JZ (2012) Supercritical fluid CO_2 extraction, simultaneous determination of total sterols in the spore lipids of *Ganoderma Lucidum* by GC-MS/SIM methods. Chem Nat Compd 48(4):657–658
38. Jin Y (2006) Study of TLC identification for sporoderm-broken spores of *Ganoderma Lucidum* capsules. Anhui Med Pharm J 10(2):117–116
39. Wang X, Chen XL, Tao JQ (2008) Identification for broken cellular wall and unbroken spore of Ganoderma lucidum by UV spectroscopy. J Anqing Teach Coll 14(3):89–90
40. Wu YK, Wang X, Wang H, Chen XL, Chen FH (2008) Identification of Anhui Jinzhai broken cellular wall and unbroken spore of Ganoderma lucidum by ^{1}H-NMR. Acta Univ Med Anhui 43(2):230–231
41. Shang XD, Li MR, Wang N, Tan Q (2005) Determination of Sporoderm-broken rate of *Ganoderma Lucidum* by hemocytometer measurement. Acta Edulis Fungi 12(2):37–40
42. Yao ZQ, Li WL, Zhou J, Zhu H, Yao HF (2007) Quality research of *Ganoderma Lucidum* spore powder. J Chin Med Mater 30(5):607–611
43. Ni WF, Xing ZT, Zhao XY, Li MR, Shang XD, Xie BG (2008) Study on determination of broken sporoderm rate of *Ganoderma Lucidum* spore product. Acta Agric Shanghai 24(3):45–49
44. Chen GZ, Liang HM, Xie YZ (2011) Quantification of triterpenoids acids in broken *Ganoderma Lucidum* spore extract by ultraviolet spectrophotometry. J Microbiol 31(1):91–94
45. Dong HL, Xia GP, Zhao NX, Bai XX, Shao ZY, Han YM (2013) Quantitative analysis of ganoderic acid C2, G, and A in sporophore and spore powder of Ganoderma by HPLC. Drugs Clin 28(1):41–43
46. Liu M, Gao L (2005) Content determination of polysaccharides of *Lucid Ganoderma* in spore powder of *Lucid Ganoderma* capsules by UV. Res Pract Chin Med 19(4):34–35
47. Qian YF, Zhang L, Pu CH (2007) Study on determination of polysaccharide in *Ganoderma Lucidum* spore powder tablets by HPLC. J Chin Med Mater 30(8):1027–1029
48. Yu Q, Shang S, Feng YL, Wang Y (2014) Determination of monosaccharide composition of polysaccharide in *Ganoderma Lucidum* spore by ion chromatography. Chin Pharm J 49(4):344–347
49. Huang XL, Wu HQ, Huang F, Lin XS (2006) Analysis of polysaccharide from broken cellular wall and unbroken spore of *Ganoderma Lucidum*. Chin Tradit Herb Drug 37(6):813–816
50. Wang WY, Chen WP (2014) Content determination of four heavy metal elements in *Ganoderma Lucidum* spores. China Pharm 17(3):508–510

51. Gao LH, Ding XM, Wang Y (2008) Hydride generation-atomic fluorescence method in the determination of arsenic in *Ganoderma Lucidum* spore. Occup Health 24(9):851–852
52. Li J, Qiu YJ, Shen GF, Zhu J (2014) Determination of 10 trace elements in *Ganoderma Lucidum* spore powder. Chin J Health Lab Technol 24(19):2767–2768
53. Gao WC, Li QY, Xian RQ, Yu HY, Wang XB (2014) Determination of residual solvents in broken cellular wall *Ganoderma Lucidum* spore powder by DI-GC. J Pharm Res 33(11):642–644
54. Lin H, Jin LZ, Che CL, Wang X, Wang YH, Wang XY (2017) Toxicological safety evaluation of sporoderm-broken spore powders of organic *Ganoderma Lucidum* of Changbai mountain. J Food Saf Qual 8(2):662–668
55. Ou QH, Lin W, Lin J (2005) Long-term toxicity of ganoderma spore powder in rats. Chin J Chin Mater Med 30(20):1784–1785
56. Qi SB, Zhou M, Li QF, Chu F (2016) Experimental study on toxicity of broken ganoderma spore powder capsules. Lab Anim Sci 33(3):27–31
57. Yang GG, Huang JK, Huang Q, Chen BF, Cai W (2007) Study on the immunomodulatory effect of ganoderma spore powder with different wall breaking methods. J Hyg Res 36(4):482–484
58. Fan QS, Hu JW, Xiao XN (2005) Studies on the immunization regulating function of grass-cultivated *Ganoderma Lucidum* spores powder. Edible Fungi China 24(5):39–41
59. Zhang RB, Chen R, Chen GM, Zheng ZW (2013) Study on effect of broken *Ganoderma Lucidum* spore powder on immune function in mice. Prev Med Trib 19(12):936–954
60. Ren W, Zou L, Zhong ZQ (2007) The effect of the *Ganoderma Lucidum* spore on the immunoregulation of the immunosuppression mice. Chin J Immunol 23:984–979
61. Zhang W, Song JK, Wang HG, Zhou QM, Wang XB, Du GH (2018) Immunomodulatory effects of wall-broken Ganoderma spore powder on immunosuppressive mice induced by cyclophosphamide. Food Drug 20(6):408–413
62. Chen JL, Han JX, Yang WY, Lu LJ, Zhang JL, Yang QL, Yuan ST, Ding J (2002) Inhibition of mixture of lucid Ganoderma and lucid Ganoderma spore on tumor cell in vitro and in vivo. Chin J Cancer 21(12):1341–1344
63. Chen C, Li P, Yao G, Xu JH (2016) Antitumor effects and mechanisms of Ganoderma extracts and spores oil. Oncol Lett 12(5):3571–3578
64. Li K, Na K, Wu K, Wang X, Wang Y (2017) Study on optimize extraction process of total triterpenoids from Sporoderm-broken spores of *Ganoderma Lucidum* and its anti-cancer effect in vitro. Chin J Mod Appl Pharm 34(9):1219–1224
65. Tan HS, Li X, Gong BL, Li G (2018) Determination of total triterpenoids in fruiting body and spores of *Ganoderma Lucidum* and assessment of their antitumor activity in vitro. Prog Microbiol Immunol 46(1):43–48
66. Li MC, Lei LS, Liang DS, Xu ZM, Yang SQ, Sun LS (2000) Effect of Ganoderma polysaccharides on interleukin 1α and tumor necrosis factor α mRNA expression in murine peritoneal macrophages. Chin J Pharmaocol Toxicol 14(3):237–240
67. Hu YH, Lin ZB, He YQ, Zhao CJ (1999) Polysaccharides isolated from mycelia of *Ganoderma Lucidum* induced on HL-60 cell apoptosis by enhancing macrophage activity. Chin Pharmacol Bull 15(1):27–30
68. Zhang QH, Lin ZB (1999) Study on antitumor activity and mechanism of Ganoderma polysaccharides B. Chin J Integr Tradit West Med 19(9):544–547
69. Zhang QH, Lin ZB (1999) The antitumor activity of *Ganoderma Lucidum* (Curt.:Fr.) P. Karst. (Lingzhi) (Aphyllophormycetideae) polysaccharides is related to tumor necrosis factor-α and interferon-γ. Int J Med Mushroom 1:207–215
70. Bao XF, Dong Q, Fang JN (2000) Structure and conformation behavior of a glucan from spores of *Ganoderma Lucidum* (Fr.) Karst. Sheng Wu Hua Xue Yu Sheng Wu Wu Li Xue Bao (Shanghai) 32(6):557–561
71. Bao X, Fang J, Li X (2001) Structural characterization and immunomodulating activity of a complex glucan from spores of *Ganoderma Lucidum*. Biosci Biotechnol Biochem 65:2384–2391

72. Bao X, Duan J, Fang X, Fang J (2001) Chemical modifications of the (1→ 3)-alpha-D-glucan from spores of *Ganoderma Lucidum* and investigation of their physicochemical properties and immunological activity. Carbohydr Res 336:127–140
73. Sun L, Zhao MY, Huang YT, Yang HY, Ma JF, Zheng ZM, Dong ZM (2006) Influence of ganoderma lucidum spore on dendritic cells from mice bone marrow and its antitumor effect. Shanxi Med J 35(8):698–700
74. Jiang Y, Wang H, Lu L, Tian GY (2005) Chemistry of polysaccharide Lzps-1 from *Ganoderma Lucidum* spore and anti-tumor activity of its total polysaccharides. Acta Pharm Sin 40(4):347–350
75. Wang PY, Zhu XL, Lin ZB (2012) Antitumor and immunomodulatory effects of polysaccharides from broken-spore of *Ganoderma Lucidum*. Front Pharmacol 3:135
76. Su JY, Su L, Li D, Shuai O, Zhang YF, Liang HJ, Jiao CW, Xu ZC, Lai Y, Xie YZ (2018) Antitumor activity of extract from the sporoderm-breaking spore of *Ganoderma Lucidum*: restoration on exhausted cytotoxic T cell with gut microbiota remodeling. Front Immunol 9:1765
77. Su JY, Li D, Chen QJ, Li MX, Su L, Luo T, Liang DL, Lai GX, Shuai O, Jiao CW, Wu QP, Xie YZ, Zhou XX (2018) Anti-breast cancer enhancement of a polysaccharide from spore of *Ganoderma Lucidum* with paclitaxel: suppression on tumor metabolism with gut microbiota reshaping. Front Microbiol 9:3099
78. Xie AM, Liu ZL, Chen L, Wu JL, Ning YP, Li BQ, Wang YK, Liu P, Li ZW (2004) Study on ultrastructural changes of substantia nigra and effects of ganoderma spores on the changes in experiment Parkinson's disease rats. Chin J Nerv Ment Dis 1:11–13
79. Zhang W, Zeng YS, Wang Y, Liu W, Cheng JJ, Chen SJ (2006) Primary study on proteomics about ganoderma lucidum spores promoting survival and axon regeneration of injured spinal motor neurons in rats. J Chin Integr Med 3:298–302
80. Zhou Y, Qu ZQ, Zeng YS, Lin YK, Li Y, Chung P, Wong R, Hägget U (2012) Neuroprotective effect of preadministration with *Ganoderma Lucidum* spore on rat hippocampus. Exp Toxicol Pathol 64:673–680
81. Qin C, Wu SQ, Chen BS, Wu XX, Qu KY, Liu JM, Zhang GF, Xu YF, Shu S, Sun L, Li YY, Zhu H, Huang L, Ma CM, Xu YH, Han YL, Lu YZ (2017) Pathological changes in APP/PS-1 transgenic mouse models of Alzheimer's disease treated with *Ganoderma Lucidum* preparation. Acta Acad Med Sin 39(4):552–561
82. Zhang ML, Bao HY, Fu WW, Yu XF, Qu SC, Sui DY (2013) Studies of cracked Ling zhi spores powder on hypnotic effects in mice. Ginseng Res 4:17–16
83. Wang WQ, Zhang ML, Yu HB, Li J, WAng XG, Wang SQ (2012) Effects of ganoderma spore powder on the expression of il-2 and intracellular Ca2+ in brain tissue of epileptic rats induced by pentaerythritol. Chin Tradit Pat Med 34(4):726–728
84. Li J, Yu HB, Yu HT, Wang SQ, Wang WQ, Jia LL (2012) Effects of ganoderma spore powder on caspase-3, NO and NOS in epileptic rats induced by pentaerythritol. Chin Tradit Pat Med 34(10):2007–2008
85. Zhang JB, Song HJ, Liu S, Wang SQ, Li XY, Lv DX, Zhai YS (2016) Study of expression of caspase-9 in pentylenetetrazol activation of hippocampal neurons by *Ganoderma Lucidum* spores. Prog Mod Biomed 16(10):1850–1853
86. Zhao HY, Li WW (2007) The study of the effects of the suppressive effects of broken wall mythic fungus spore powder on liver fibrosis on cell factor. Chin Pract Med 5:12–14
87. Piao RL, Cui Y, Xu QS (2011) Protective effects of the oil from spores of *Ganoderma Lucidum* on carbon tetrachloride-induced hepatic injury in mice. J Med Sci Yanbian Univ 1:20–22
88. Hu ZM, Zhou YL, Deng YY, Li JP (2016) Protective effect of *Ganoderma Lucidum* spore powder on CCl_4 induced liver fibrosis in mice. Cent S Pharm 14(7):696–699
89. Yang LL, Deng ZQ, Huang WY, Chen LX, Zhu LY (2013) Protective effect of *Ganoderma Lucidum* against sub-acute alcoholic hepatic injury in mice. Lishizhen Med Mater Med Res 24(3):513–514

90. Zhao YP, Feng CH, Fu CL, Hua HY (2017) Effect of broken *Ganoderma Lucidum* spore powder on serum ALT, AST levels and liver inflammation in mice with ConA-induced immune injury. Zhejiang J IntegrTradit Chin West Med 9:760–762
91. Gu X, Wang Z, Liu GT (1993) Pharmacology studies on the spores of *Ganoderma Lucidum* II. Effect on lipid peroxidation of skeleton muscle homogenate and on generation of superoxide anion. Pharmacol Clin Chin Mater Med 9:9–12
92. Zhang SQ, Shi XJ (2003) Experimental study on cardiac protection and anti-oxidation effects of Jisheng injection. Chin J Integr Tradit West Med 6:455–457
93. Liu Y, Wang SQ, Kang YM, Ma XR, Zhang B, Zhang T, Liu JX, Liang YF, Wang FF (2008) Protective effect of *Ganoderma Lucidum* spores on mitochondrial oxygen free radical damage in kidney of type 2 diabetic rats. Chin J Gerontol 7:634–636
94. Hu ZM, Liu JN, Zhou YL, Liu JP (2016) Lingzhi Baozifen (LZBF) attenuates stomach inflammation induced by ethanol in mice. Shaanxi J Tradit Chin Med 5:632–634
95. Tong GH, Ma L, Wang HM (2004) Studies of *G. lucidum* spores powder on the immunoregulation and antiradiation. Chin J Food Hyg 2:132–134
96. Yan Y, Xu JH, Yang F (2004) Research on the radioprotective effect of *Ganoderma lucidum*. Chin J Food Hyg 5:429–430
97. Chen R, Lin J, Chen GZ (2007) Study on radiation resistance of *Ganoderma lucidum* spore powder. Strait J Prev Med 2:53–54
98. Zhang X, Cui WM, Liu ZQ, Wang W, Li N (2003) Study on antimutagenic and antineoplastic effect of *Ganoderma Lucidum* spore powder. Chin J Public Health 2:173–174
99. Wei LZ, Xu DX (2000) Studies on analgesic and anti mutagenic effects of spore of *Ganoderma Lucidum*. Chin J Public Health 8:700–701
100. Qi YF, Li XR, Yan M, Liu A, Jiao ZH, Liu Y (1999) Clinical observation of adjuvant chemotherapy by *Ganoderma Lucidum* spore for the treatment of digestive tract cancer patients. Chin J Integr Tradit West Med 19(9):554–555
101. Zhen ZJ, Chen YJ, Yong JI, Wang FJ, Li JY, Li QH, Wu ZP, Fan GY, Chen HW (2012) Inhibitory efficacy of *Ganoderma lucidum* spore on recurrence after curative resection of hepatocellular carcinoma. J Dig Oncol (Electron Version) 4(1):40–43
102. Zhen ZJ, Wang FJ, Fan GY, Chen HW (2013) Effect of ganoderma lucidum spore to the immunological function of patients with hepatocellular carcinoma after operation. Chin J Hepat Surg (Electron Ed) 2(3):171–174
103. Wang J, Chen ZQ, Li Y, Wang XS, Fei CL, Wang SJ (2016) Clinical efficacy of compound *Ganoderma Lucidum* spore capsule combined with chemotherapy on non-small cell lung cancer and its effect on immune function. Mod J Integr Tradit Chin West Med 25(24):3762–3764
104. Zhao H, Zhang Q, Zhao L, Huang X, Wang J, Kang X (2012) Spore powder of *Ganoderma Lucidum* improves cancer-related fatigue in breast cancer patients undergoing endocrine therapy: a pilot clinical trial. Evid Based Complement Alternat Med 2012:809614. https://doi.org/10.1155/2012/809614

Chapter 7
Development and Innovation of Ganoderma Industry and Products in China

Zhenhao Li, Jianlong Zhou, and Zhibin Lin

Abstract Ganoderma (Lingzhi) has been used as a medicinal mushroom to promote health in China for more than 2000 years. The modern research and development of Ganoderma industry started from the 1950s, in which the successful cultivation of Ganoderma fruiting body and submerged fermentation of Ganoderma mycelium lay the critical foundation for the industry development. Recent decades have witnessed the rapid development of Ganoderma industry, which is boosted through various efforts made by the government, the academia, and the industry. In this chapter, the development of Ganoderma industry in China is reviewed in terms of gross output, standards, scientific articles, patents, and associations. In addition, development of Ganoderma products and manufacturing technologies are also overviewed and summarized. In the last section, several innovation trends are suggested for the further development of Ganoderma industry.

Keywords Ganoderma · Industry · Product · Development direction

7.1 Overview of Ganoderma Industry in China

Ganoderma (Lingzhi) has been used as a medicinal mushroom to promote health and prolong lifespan in China for more than 2000 years. In ancient times, Ganoderma was mainly collected from wild resources, and artificial cultivation of this fungus was rarely recorded. This made Ganoderma one of the most precious traditional medicines, while on the other hand hampered the broad use and development of

Z. Li (✉)
Zhejiang Shouxiangu Pharmaceutical Co., Ltd, Wuyi, Zhejiang, China

College of Pharmaceutical Sciences, Zhejiang University, Hangzhou, Zhejiang, China

J. Zhou
Zhejiang Shouxiangu Pharmaceutical Co., Ltd, Wuyi, Zhejiang, China

Z. Lin
Department of Pharmacology, School of Basic Medical Sciences, Peking University, Beijing, China
e-mail: linzb1937@sina.com

Z. Lin, B. Yang (eds.), *Ganoderma and Health*, Advances in Experimental Medicine and Biology 1181, https://doi.org/10.1007/978-981-13-9867-4_7

Ganoderma during the old times. As a fact, although Ganoderma was considered as an "immortal herb" in traditional Chinese medicines (TCM), this precious medicine was not widely used in ancient herbal formulae due to lack of the resource. In the 1950s, Ganoderma was successfully cultivated by researchers from Institute of Microbiology, Chinese Academy of Sciences [1]. Since then, artificial cultivation of Ganoderma fruiting body and submerged fermentation of Ganoderma mycelia were gradually popularized in China, which lay down the critical foundation for product and industry development of Ganoderma. In the 1970s, based on the previous chemical and pharmacological studies, clinical researches of Ganoderma were conducted which preliminarily demonstrated the therapeutic efficacy of Ganoderma [1]. At the same time, commercialization of Ganoderma began to emerge with the increase of cultivation areas and the development of related products [2]. In the 1970s, Ganoderma was widely produced and used as a medicine in China [3]. China's reform and opening up also accelerated the development of Ganoderma industry. In view of its satisfactory effect and medical potential in disease treatment and prevention, many Ganoderma products, including drugs, supplements, cosmetics, and foods, were developed since the 1980s [4]. The increased consumption of Ganoderma materials led to the expansion of cultivation area, as well as production and manufacturing scale, which in turn improved related technologies [2, 5]. Since then, the development of Ganoderma industry entered a virtuous cycle with a high growth rate. In 2015, the global annual gross output of Ganoderma industry was estimated to be worth US$5.0 billion with fruiting body and spore yields of one hundred and sixty thousand tons, of which China accounted for 30% in the gross output and 75% in yields [6].

With the development of Ganoderma market, standards for ensuring the safety and effectiveness are much sought after by industry and other stakeholders such as consumer groups. Since 2000, the *Pharmacopoeia of the People's Republic of China* (Part 1) have listed the fruiting body of *Ganoderma lucidum* (Leyss. ex Fr.) Karst (Chi Zhi) and *Ganoderma sinensis* Zhao, Xu et Zhang (Zi Zhi) as Lingzhi, which clarify the legal status of Ganoderma. Besides, *United States Pharmacopeia, Herbal Medicines Compendium*, and the *Korean Herbal Pharmacopoeia* have also included *G. lucidum* that specify quality requirements and test methods. In China, more than 50 national, industry, provincial, and group standards of Ganoderma have been developed concerning collection, cultivation, production, identification, quality, and strain. A milestone of Ganoderma standard development is the publication of the international standard of *Ganoderma lucidum* by ISO/TC 249 (International Organization for Standardization/Technical Committee 249 Traditional Chinese Medicine) in December 2018, which was first proposed by China and jointly developed by other seven member states including Australia, Canada, Germany, Republic of Korea, South Africa, Thailand, and USA [7]. This ISO standard not only helps to ensure the safety and quality of global Ganoderma-related products and market but also functions as an international consensus for better regulation of the international trading. Details of quality requirements of the representative Ganoderma standards are shown in Tables 7.1 and 7.2.

The development of the Ganoderma industry also boosts the academic researches. We collected the publications on Ganoderma between 1998 and 2018 from Scopus

Table 7.1 Limits of moisture, total ash, water-soluble extractives, polysaccharides, and ganoderic acid A in *Ganoderma lucidum* fruiting body in different standards

Authority regulation	Moisture	Total ash	Water-soluble extractives	Polysaccharides	Ganoderic acid A
Chinese Pharmacopoeia 2015 ed.[a]	NMT 17%	NMT 3.2%	NLT 3.0%	NLT 0.9%[b]	–[c]
United States Pharmacopeia, Herbal Medicines Compendium 1.0	NMT 17%	NMT 4.0%	NLT 3.0%	NLT 0.7%[d]	NLT 0.3%[e]
The Korean Herbal Pharmacopoeia, English Edition	–	NMT 1.5%	–	–	–
ISO 21315: 2018	NMT 17%[f]	NMT 4.0%[f]	NLT 3.0%[f]	Determined[g]	Determined[g]

[a]In Chinese Pharmacopoeia, *Ganoderma lucidum* and *Ganoderma sinensis* are included in GANODERMA
[b]Determined by ultraviolet-visible spectrophotometer using glucose as reference standard
[c]"–" means the index is not set in the pharmacopoeia
[d]Determined by HPLC using five reference standards (mannose, D-glucuronic acid, galactose RS, dextrose, L-fucose), sum of which are calculated as the content of polysaccharides
[e]Determined by UHPLC, and ganoderic acid A is used to calculate the content of total ten triterpene acids by conversion factor method, sum of which are calculated as the content of triterpene acids
[f]The value can be modified by the National Standards Bodies
[g]No limit value is specified

Table 7.2 Limits of heavy metals in *Ganoderma lucidum* fruiting body in different standards

Authority regulation	Arsenic	Cadmium	Lead	Mercury
Chinese Pharmacopoeia 2015 edition	–	–	–	–
United States Pharmacopeia, Herbal Medicines Compendium 1.0	NMT 3.0 μg/g	NMT 0.5 μg/g	NMT 5.0 μg/g	NMT 0.2 μg/g
The Korean Herbal Pharmacopoeia, English Edition	–	–	–	–
ISO 21315: 2018	Determined	Determined	Determined	Determined

and CNKI databases using "Ganoderma," "Lingzhi," or "Reishi" as the keywords. As shown in Fig. 7.1, the number of annual publications increased rapidly from 1998 to 2018 with approximately 800 articles published per year in recent years. In addition, the number of papers in Scopus, most of which are English-language studies, shows a rapid increase. This indicates that Ganoderma has attracted global interests, and the depth and width of Ganoderma-related studies also reach a relatively high level. Actually, in addition to China, the USA, South Korea, Malaysia, and Brazil also have a certain number of article outputs of Ganoderma every year.

The categories of Ganoderma papers are further analyzed. Biochemistry, molecular biology, agricultural and biological sciences, medicine, pharmacology, toxicology and pharmaceutics, and chemistry represent the most abundant categories with

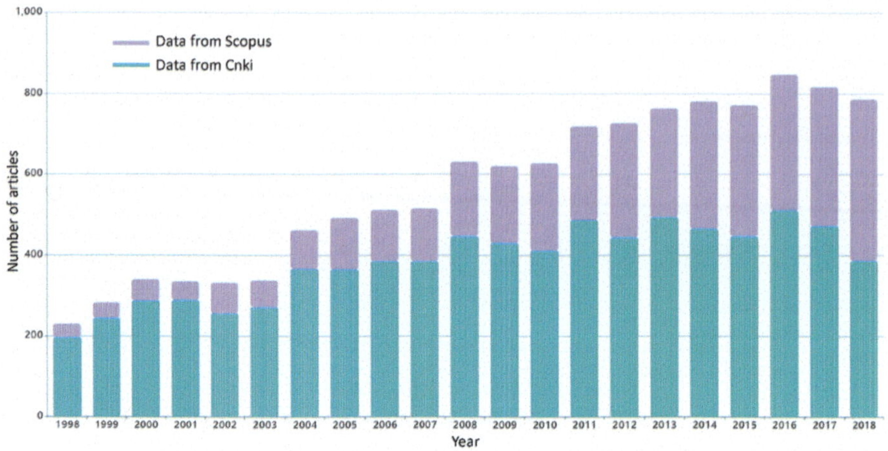

Fig. 7.1 The number of published articles of Ganoderma during 1998–2018

higher frequency. Moreover, we find that pharmacological studies of Ganoderma lie in several specific subjects, such as immunoregulation, cancer treatment, antioxidation, cardiovascular diseases, diabetes, liver protection, and neuropharmacology. In fact, cancer immunotherapy has emerged one of the most popular fields of Ganoderma research, suggesting a promising direction for new drug development of Ganoderma.

It is worth noting that with the development of the industry, new medicinal parts of Ganoderma, such as mycelia and spores, have been developed, as evidenced by the increasing number of related articles. Traditionally, Ganoderma is cultivated on the cut-log or the bulk bag [6, 8], and most of bioactivities are found in the fruiting bodies of Ganoderma [9]. However, such cultivation methods require much resource and take long time for harvest. To overcome such limitations and meet the increasing commercial demand, submerged fermentation technique of Ganoderma mycelia develops rapidly, which only takes weeks to harvest with higher yields and lower cost. In addition, compared with fruiting body cultivation, submerged fermentation provides more stable culture conditions to maintain the quality of mycelia of different batches and is more compatible to modern manufacturing equipment. However, chemical compositions of Ganoderma mycelia are different from those of the fruiting body, indicating their potential difference in health benefits [10]. Generally, the fruiting body is considered to possess better efficacy than the mycelia do [11].

Compared with scientific publications, patents provide more valuable information from an industry perspective. Therefore, we performed a statistical analysis of the Ganoderma patents applied in China from 1998 to 2018. As depicted in Fig. 7.2, patent applications of Ganoderma shows a rapid increase with a climax of approximately 2700 applications in 2016. It indicates that the research, development, and innovation of Ganoderma grow rapidly, and more and more attentions have been paid to intellectual property. However, both the patent applications and granted patents showed a decreased trend since 2016, as the State Intellectual Property Office began

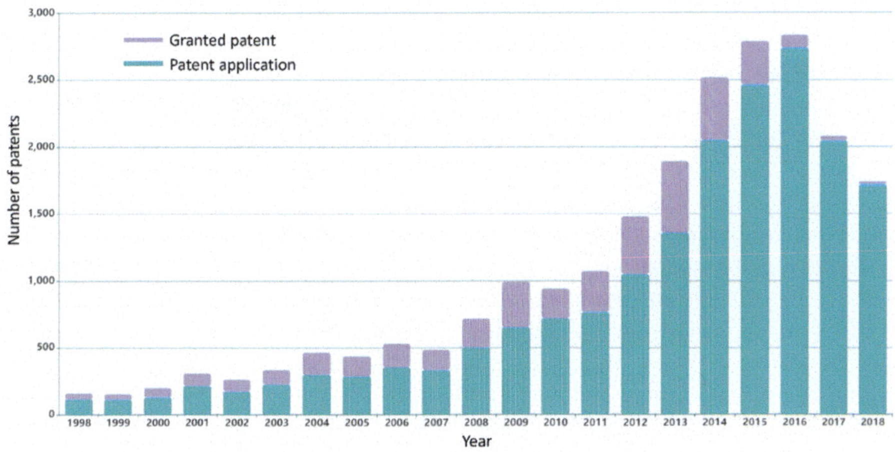

Fig. 7.2 The number of patents of Ganoderma applied in China during 1998–2018

implementing new policies to limit unqualified patent applications. Applicants of the patents, including companies, institutes, universities, and individuals, from Anhui, Guangdong, Jiangsu, Shandong, Zhejiang, and Shanghai account for the major proportion (>50%). Further analysis of the patents reveals that a plenty of the patents are related to medicines (e.g., active components, drugs), other products (e.g., supplements, foods), cultivation and manufacturing methods, and gene sequences of Ganoderma. Active fractions, components, and compounds separated from fruiting bodies, spores, spore oils, and mycelia of Ganoderma are widely patented for disease treatment and prevention, including tumors, hepatitis, coronary artery diseases, and neurodegenerative diseases. An interesting observation is that individual applicants apply for more patents related to combinations of Ganoderma with other medicinal herbs, while companies, institutes, and universities focus on bioactive components and medicinal use of Ganoderma. A possible reason for this fact is that it is relatively easy to invent herb combinations containing Ganoderma, while identification of bioactive components requires more researches and investments.

Companies are the leading force for industry development. According to incomplete statistics, there are more than 300 companies in China whose major businesses are related with Ganoderma in terms of cultivation, manufacturing, extractives, and drug/product development [12]. Prominent brands of Ganoderma products are emerging such as Shouxiangu, Ganoherb, Zhongke, Yuewei, Xiankelai, Zhenqin, Anhui, and so on. Some of larger-sized Ganoderma enterprises have established the so-called whole industry chain operational structure including new cultivar breeding, cultivation, processing, manufacturing, product development, sales, and post-market research of Ganoderma. Such a holistic development approach not only ensures the safety and quality of raw materials and drug products but also is helpful to supply-chain's quality and reliability, the lack of which is often attributable to quality issues [13]. Many standards and guidelines, such as good agricultural and collection practices (GACP), good manufacturing practice (GMP), and organic certification, are gradually applied in cultivation and production processes to obtain materials and products of good quality. The application of these standards promotes sustainable use of Ganoderma. In addition, large-sized companies actively employ state-of-the-art technologies for the modernization of the industry. For example, some companies have utilized modern agricultural, engineering, and internet of things (IOT) technologies to precisely control the cultivation and manufacturing processes, through which safety, efficacy, and consistency of Ganoderma products can be better guaranteed [6, 14].

Professional associations, committees, and academic meetings play vital roles in sci-tech exchange and the integration of production, teaching, and research. The 1991 National Symposium on Ganoderma and the 1994 International Symposium on Ganoderma held in Beijing by Chinese Pharmaceutical Association, China Edible Fungi Association, and Beijing Medical University (i.e., Peking University Health Science Center, PUHSC) initiated the exchange of industry and academic community, as well as international collaborations [1]. Moreover, the International Meeting on Ganoderma Research held annually by International Society of Ganoderma Research, Chinese Pharmacological Association, China Edible Fungi Association, and their sponsors has been a very important forum for Ganoderma research, which contributes significantly to the development of the Ganoderma industry. It is worth mentioning that in 2018, approved by China Association of Traditional Chinese Medicine, the Ganoderma Committee was founded, and the inaugural meeting was held in Wuyi, Zhejiang Province, in May 5–6, 2018. Experts, researchers, and practitioners from different organizations and companies all over the world attended the meeting. Several international collaborations were discussed and eventually achieved during this meeting. This is a milestone of Ganoderma industry in China, which holds great significance for the industrial development.

7.2 Ganoderma Products in China

As demonstrated in modern biological research, Ganoderma possesses numerous functions such as immunity enhancement, lowering blood glucose, liver protection against chemical injuries, anti-radiation, antitumor, anti-aging, anti-inflammation, anti-blood clotting, and relief of insomnia [8]. Thus, applied research and product development have been performed in all areas of the Ganoderma industry chain from basic research to applications, from new cultivar breeding, large-scale standardized plantation, processing, manufacturing, to quality control. In China, over 150 research teams or groups are engaged in Ganoderma scientific and technological studies, and about 300 enterprises are engaged in applied research, product development, and/or product sales. Approximately 180 medicines, 1200 health foods, and 3300 cosmetic products containing Ganoderma compositions have been manufactured and promoted to the market, including herbal medicines, nutraceuticals, health foods, dietary supplements, cosmetics, ornaments, and agricultural products. The diverse Ganoderma products currently on the market are made from fruiting bodies, spores, mycelia, and fermentation liquid through simple or sophisticated processes. Ganoderma made into a final product either itself or in combination with other functional ingredients and in the forms of fruit body dry slice, fine powder, tea drink, alcohol drink, beverage, capsule, spore powder, broken-wall spore powder, spore oil, extract-paste, mycelia powder, tablet, granule, cream, and others through processing such as cutting, drying, grinding, breaking, extraction, mixing, and homogenizing [12]. The Ganoderma products are well accepted by customers domestically and overseas, as the health functions of Ganoderma are mostly proved substantial and reproducible among general population.

7.2.1 Ganoderma as Medicines

At China National Medical Product Administration's (NMPA's) website, keyword search using Lingzhi or Zizhi (i.e., *Ganoderma lucidum* or *Ganoderma sinensis*) revealed that there are 178 registered drugs containing *Ganoderma*. Among them are capsule 81, pill 41, granule 25, syrup 12, vina (alcoholic drink) 9, oral liquid 6, tea bag 2, dripping pill 1, and injection 1. *G. lucidum* is frequently formulated with other traditional Chinese medicine (TCM) materials as compound medicines for tonifying liver and kidney, tranquilizing heart and mind, invigorating spleen and removing dampness, and benefiting stomach and blood. In other words, *G. lucidum* is used for treatment of chronic bronchitis, inflammation, insomnia, coronary heart disease, neurasthenia, hepatitis, chronic hepatitis B, stomach ulcer, hyperlipidemia, tumor, etc. The *Pharmacopoeia of the People's Republic of China* (Part 1) (2015 Edition) records nine Ganoderma-containing formulations in the forms of sliced decoction pieces, prescription preparations, and mono-herbal preparations.

Many formulations of Ganoderma compound medicines have inadequate scientific support to explain why certain raw materials are included. The chemical composition of compound herbal medicines is quite complex, and the safety of such medicines is not well understood. Most of Ganoderma medicines currently lack clear characteristic compounds that can be quantitatively measured for each formulation and they do not have unique chemical profiles to distinguish among themselves. The quality standard in general is not sufficiently detailed and not up to the criteria of modern medicines. Although it has long been used in history as remedies for various illnesses, Ganoderma has yet to become a major drug either on its own or in combination with other herbs as a compound medicine. However, with extensive and intensive research performed on this herb in recent years, it will not be far away for Ganoderma to develop into an effective modern drug.

7.2.2 Ganoderma as Health Food

In addition to its usefulness as medicines, Ganoderma is formulated in numerous health foods (functional foods or nutraceuticals or dietary supplements) for improving human health and promote longevity. According to China NMPA's website, there are 1233 Ganoderma-containing products currently on file, including 1215 domestically made and 18 imported products. They are in the form as capsules, tablets, granules, oral liquid, tea bags, paste, vina, etc. The manufacturing process usually includes extraction of Ganoderma fruiting bodies or spores by water or alcohol after crushing and grinding of raw materials, then concentration and drying, and then made into various forms of Ganoderma health products according to designed formulations. Legally, each Ganoderma health food product, before sold on the market, must be successfully registered with Chinese government and has to specify the daily intake dosage and amount of active ingredients such as triterpenoids, polysaccharides, and oleic acid. In this regard, Ganoderma health food products have clearer, more objective, and measurable indicators, and thus their product quality can be guaranteed.

Among these health food products, capsule is the most common and simplest preparation. Approximately 70% of Ganoderma health products were made into capsules, followed by granules and tablets. Common raw materials for making Ganoderma health products are fruit body extract, fruit body fine powder, mycelia, spores, and spore oil. They are formulated either individually or in combination with other bioactive natural products. Analysis of these Ganoderma health food's compositions showed that products of Ganoderma-only, with one, two, three, or four and more other raw materials, account for ~43.5%, 23.0%, 7.5%, 15.5%, and 10.5% of the Ganoderma health products, respectively.

Ganoderma extract is one of the most frequently used components in health food. China NMPA allows health foods to legally claim 27 functions, and Ganoderma extract-containing health foods generally claim functions including immunity enhancement, insomnia improvement, auxiliary protection against chemical-induced liver injury and radiation-caused damage, fatigue alleviation, auxiliary lowering of blood lipid and blood glucose and etc. Ganoderma extract is usually formulated with Ganoderma spores, ginseng, other herbs, even black rice powder, etc. In general, the formulation rule for designing herb-based health food or compound medicines is Monarch-Minister-Assistant-Guide (or pronounced jūn chén zuǒ shǐ in Chinese), wherein monarch is the component targeting a main symptom, minister is the component helping monarch's function, assistant targeting secondary symptoms, and guide leading the monarch, minister, and assistant to where the symptoms (or causes) are or modulating the side effect caused by the monarch, minister, and assistant.

7.2.3 Ganoderma as Cosmetic Products

With deeper understanding of bioactive compositions in Ganoderma, many molecules are known to be absorbed by cells to regulate the metabolism of skin cells. Thus the whole extract or partially purified extract of Ganoderma (*G. lucidum,*

G. sinensis), as legally permitted for use as cosmetic raw material, are currently employed as ingredients in manufacturing a diverse array of cosmetic products. According to China NMPA's website, there were over 3000 cosmetic products that contain the extract of Ganoderma. Currently, only about 350 such cosmetic products are active on the market.

Ganoderma cosmetics appear mostly as sunblock cream, skin repair cream, skin de-freckle cream, facial mask, essence milk, foundation liquid or powder, shampoo, body lotion, etc.

7.2.4 Other Products of Ganoderma

Often it is the extract (intracellular composition) of Ganoderma that is used in medicines, cosmetics, and foods. The large amount of solid residue produced from extraction of Ganoderma is mostly underutilized and consequentially presents a major environmental problem. How to utilize the residue ecologically needs creativity.

Chitin $(C_8H_{13}O_5N)_n$ is one of the chemicals found in the residue of Ganoderma, which is of N-acetylglucosamine through beta-1,4 covalent bond of 1000–3000 units. Chitin is degraded enzymatically or by strong acids, releasing the acetyl group to become chitosan and monosaccharide. Chitin can inhibit growth of a broad range of bacteria, enhance immunity, and inhibit spread of tumor cells [15]. Moreover, this polysaccharide can also decrease blood cholesterol and blood lipids, reducing the risk for high blood pressure [16]. Chitin and its deacetylation product chitosan have been used in medicines, cosmetics, food additives, animal feeds, and so on [15]. For example, chitin is not allergic to humans; thus it can be made into tough and strong surgical string or membrane patch for wound sealing and healing. The biomaterial sacchachitin prepared from the fruiting body of *Ganoderma tsugae* can help and accelerate wound healing and can be used as synthetic contact lenses, artificial skin, artificial dialysis membranes, and artificial blood vessel and for treating corneal trauma and bacterial acne [17]. Chitin can activate cell growth, stimulate repair of cell damage, and help keep skin moist; thus it can be used in cosmetic products as a functional agent for moisture and for stimulating skin cell regeneration [18]. Therefore, it is highly valuable to purify chitin and chitosan from the residue of either fruiting body or spores to enhance the value of Ganoderma.

Ganoderma is a unique auspicious thing in Chinese history. It is considered as a symbol of heaven's will, beauty, auspiciousness, wealth, and longevity. Ganoderma not only has medical and health benefits but also has important cultural appreciation value (detail see Chap. 1). Ganoderma bonsai (or miniascape), with its unique classical artistic shape and profound artistic connotation, has become the first choice for decorating high-end rooms, offices, and hotels, as well as the fashion of gift and collection. Moreover, Ganoderma jade accessories and Ganoderma porcelain are very popular with consumers at home and abroad for their rich cultural connotations and beautiful implications.

7.3 Manufacturing Technology of Ganoderma Products

In recent years, many new and advanced manufacturing methods have been proposed for manufacturing of traditional Chinese medicines [19], as well as for Ganoderma products, though some of which are still at a laboratory scale. Therefore, for new drug and product development, the producers are suggested to identify which components of Ganoderma are effective for target diseases and to evaluate these new methods for real-world applications, investigate the scale-up approaches, and then develop the manufacturing methods accordingly.

We therefore review the extraction/purification procedure of polysaccharides and terpenoids in the publications of recent years. We think these cutting-edge technologies would be helpful for developing manufacturing procedures of modern Ganoderma products.

Polysaccharides are considered as important contributors to the bioactivities of Ganoderma [20]. Different kinds of polysaccharides including polysaccharides, glycopeptides, polysaccharide-protein or polysaccharide-peptide complexes have been extracted and isolated from the fruiting bodies or spores of Ganoderma [20]. Therapeutic applications such as antioxidant, anticancer, immunomodulator, and antibacterial have been reported for Ganoderma polysaccharides [20, 21]. The basic units of Ganoderma polysaccharides include glucose, fucose, xylose, arabinose, mannose, and galactose, which are linked by different types of glycosidic bonds and can be bound to proteins or peptides [22]. As different Ganoderma polysaccharides possess different physical and chemical characteristics, it is very important to select appropriate extraction and purification methods. The most widely used extraction solutions for Ganoderma polysaccharides include hot water or acidic, saline, and dilute alkali solutions or with dimethyl sulfoxide [20, 23]. Hot water extraction is the most common method for extraction of herbal medicines including Ganoderma because it is cost-effective, safe, and environment friendly. High temperature is helpful to release active constituents inside the cell walls. In order to remove accompanying low-molecular compounds and lipids, raw materials of Ganoderma are often extracted by organic solvents (e.g., 95% aqueous ethanol) first, followed by extractions with hot water or saline/diluted alkali to yield crude water-soluble or water-insoluble polysaccharides, respectively [24]. Another widely used approach is hot water extraction followed by ethanol precipitation [25]. The extraction procedure is usually combined with other methods such as microwave, ultrasonic, and enzymatic method to increase the yield of polysaccharides [23]. After extraction, free proteins need to be removed by Sevag, ultrafiltration, dialysis, or other methods [20, 25]. Then, to further purify the polysaccharides, freeze-thawing method and different chromatographic methods, such as ion-exchange, gel filtration, and affinity chromatography, are used to obtain the pure polysaccharides [20, 26].

Besides polysaccharides, Ganoderma triterpenes are well considered as bioactive metabolites to a wide range of medical effects [11, 27]. Triterpene is composed of six isoprenes to form a four- or five-membered ring structure, and more than 316 triterpenes have been identified from Ganoderma till now [27]. The water solubility

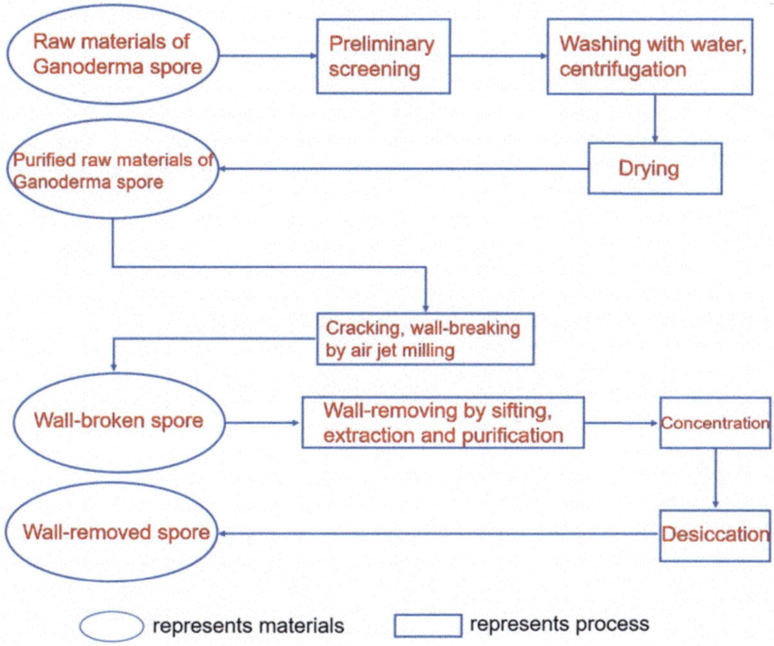

Fig. 7.3 Manufacturing process of the wall-removed Ganoderma spore of the Chinese patent CN201310743712.8

of triterpenes is very slight, so raw materials of Ganoderma are generally extracted by organic solvents first (e.g., 95% aqueous ethanol) with reflux, microwave, and ultrasonic method [28]. Another feasible method is extraction using supercritical fluid carbon dioxide (CO_2) which is more convenient and green with relatively low energy consumption [29]. Similar to polysaccharides purification, chromatographic methods including silica gel and macroporous resin are used to enrich the triterpenes after extraction [28]. It is worth noting that solvents used for real production, especially supplements and foods, should be carefully considered and selected. Besides, most of these extraction and purification methods for polysaccharides and triterpenes are in the lab or pilot scale, thus engineering and process validation batches are necessary to validate and optimize the manufacturing processes [30].

With the development of Ganoderma researches, manufacturing technologies, and equipments, more and more advanced techniques have been employed to obtain high-quality Ganoderma products. For example, a Chinese patent CN201310743712.8 provides a comprehensive method for the production of the so-called wall-removed *Ganoderma lucidum* spore. As shown in the Fig. 7.3, this method employs many state-of-the-art techniques to screen better quality raw materials of *Ganoderma lucidum* spore, break the sporoderm of the spore by supersonic air jet milling, and finally remove the sporoderm by solvent extraction and filtration. As a result, the contents of active components such as polysaccharides and triterpenes are much higher than other Ganoderma spore products. Several pharmacological studies shown in the patent also demonstrate that such a high-quality product exhibits better therapeutic effects in tumor treatment and immuno-enhancement when compared with general wall-broken Ganoderma spore. Moreover, this patent provides the comprehensive utilization of the resulting sporoderm in the application of antioxidant, anti-lipid peroxidation, and digestion improvement, which is an example for the resources recycling utilization of Ganoderma.

7.4 Innovation Trend of Ganoderma Industry

7.4.1 New Drug Discovery and Development of Ganoderma

Herbal medicines have been a very important source for new drug discovery and development. About half of new chemical entities approved by the US Food and Drug Administration (FDA) from 1981 to 2014 are derived from natural products [31]. Moreover, there is a renewed interest in the discovery and development of novel therapeutic agents from natural products and herbal medicines. Ganoderma has received great attention for new drug development owing to the ample scientific evidences for its health benefits, and several clinical trials are already ongoing in treatment of tumors, Gulf War illness, and eczema [11]. Considering the illustrious efficient pharmacological properties of Ganoderma and its metabolites such as terpenoids and polysaccharides, it is possible to evolve this ancient medicine to therapeutic agents in the form of either new chemical entities (or active pharmaceutical ingredients) or botanical drugs. However, to fulfill this ambitious goal, more efforts should be done to clarify the active components and mechanism of actions, and rigorously designed clinical trials are required to confirm the safety and clinical efficacy, as well as strict quality control metrics and manufacturing process to ensure drug quality.

A basic but very important part in drug development process is the breeding of novel cultivars having much higher content of active components such as polysaccharides and triterpenoids. New breeding technologies including genetic engineering and synthetic biology are expected to be used to create novel Ganoderma varieties producing higher level of a particular terpene or a group of triterpenoids, thereby providing a sufficient supply of raw material to developing active components into medicines or functional foods. Thus, it requires deep understanding of signal transduction pathways and biosynthetic genes, which control the secondary metabolite biosynthesis in Ganoderma.

The identification of lead compounds from Ganoderma specific to certain disease targets is a classic, tough but still promising path for drug discovery. Actually, several compounds isolated from Ganoderma exhibit potent activities to different diseases. For example, Lingzhi-8 is an immunomodulatory protein isolated from

Ganoderma lucidum, consisting mainly of 110 amino acids and having an immunoglobulin-like structure [32]. Lingzhi-8 exerts potent therapeutic efficacies on lung cancers and gastric cancer by interfering with cell adhesion and focal adhesion kinase (FAK) functions [33] and targeting epidermal growth factor receptor (EGFR)-dependent processes [34]. 7d is a newly reported dual inhibitor of HMGCR and α-glucosidase, which is a derivative of Ganomycin I isolated from Ganoderma [35, 36]. This compound can improve metabolic dysfunctions in obese mice through modulating the gut microbiota [36], which is a promising drug candidate for metabolic syndrome. Su and her colleagues found that Ganoderma triterpenes could attenuate cyst development in vivo, and CBLZ-7 (ethyl ganoderate C2) had a potent inhibitory effect on cyst development in vitro, which may be potential therapeutic reagents for treating autosomal dominant polycystic kidney disease [37]. These drug candidates hold significant promise for new drug development, while it should be noted that the contents of these compounds in raw materials of Ganoderma are generally very low; thus direct purification would be labor-intensive and costly. Methods such as recombination, synthesis, semi-synthesis, or biosynthesis can be taken into consideration for industrial production.

In addition to new chemical entities, the development of Ganoderma drug products in the context of botanical drugs is another feasible way for new drug development. Compared to highly purified drugs, botanical drugs exert their effects through the interference of multiple constituents on various biological targets, thus holding significant promise for the prevention and treatment of complex diseases [38]. Botanical drugs of Ganoderma could derive from a single raw material of Ganoderma or from multiple botanical raw materials containing Ganoderma (generally called herb compatibility in traditional Chinese medicine). According to current studies of Ganoderma, one possible candidate can be Ganoderma triterpenes, polysaccharides (including polysaccharides, glycopeptides, polysaccharide-protein, or polysaccharide-peptide complexes), or combinations thereof. Ganoderma triterpenes and polysaccharides can play antitumor roles through different molecular targets and pathways [21] (e.g., immunomodulator, cytotoxic effect, prevention of metastatic growth, inducement of cell cycle arrest), thus enhancing the anticancer effects and alleviating the side effects. Moreover, as triterpenes are generally lipophilic while polysaccharides are hydrophilic, the extraction and production processes of Ganoderma triterpenes and polysaccharides can be compatible, which would take full advantage of raw materials. Polyherbal preparations containing Ganoderma are also possible research and development direction. According to traditional Chinese medicine (TCM) theory, different herbs can be properly combined to produce compatible (synergistic) effects while preventing incompatible (antagonistic) influences for disease treatment [39]. Such combinations can be derived from clinical formulae prescribed by the TCM physicians, which already have preliminary clinical evidence. However, for such multi-herb drugs, both nonclinical and clinical studies should be carefully designed and conducted, and the past human experience should be thoroughly reviewed, thereby increasing the success rate of drug development. In recent years, a new TCM drug development model called component-based Chinese medicine has been developed [40], which includes isolation of fractions of medicinal herbs or traditional remedy, identification of active components, and combination of active components using multiple state-of-the-art technologies such as in silico modeling, large-scale screening, and rational design based on knowledge from disease-target networks [40]. This novel methodology for multi-component drug design can also inspire the drug development of Ganoderma, which has the potential to improve the efficacy, reduce the dosage, and increase patient compliance.

In recent years, the success rate for developing a new drug is showing a decreasing trend for chemical drugs, botanical drugs, and Chinese patent medicines [41–43]. In order to improve the success rate for Ganoderma drug development, clinical indications, pharmacological and toxicological studies, and chemistry, manufacturing, and controls (CMC) should be carefully designed and studied. As described in Sect. 7.1, Ganoderma-related studies focus mainly on molecular biology, and pharmacology and medicine, while CMC researches are largely ignored and underestimated. CMC is generally used by US FDA to describe the data for the nature, manufacturing process, and testing of the drug substance and drug product. This term is often interchangeably used with the ICH term-Module 3 Quality. Generally, the CMC aspects will cover the characterization of the active substance, dosage form, raw materials, description of the product, process development, and manufacturing processes, in-process control, analytical controls, and so on [44]. For Ganoderma drug product development, works can be conducted in the following parts: (1) qualitatively and quantitatively analyze the chemical compositions of the Ganoderma drug product and correlation analysis with the data from TCM clinical applications and biological studies, on such bases to ascertain the active components and molecular entity of the drug product; (2) perform different animal experiments to understand the safety and mechanism of actions of the drug product and thus clarify the quality profile; (3) perform clinical trials to demonstrate the efficacy and safety; (4) bridge clinical batches well with pilot batches, engineering batches, process validation batches, and commercial batches; and (5) develop appropriate quality control metrics including in-process control, quality testing methods, and quality standards, to make sure the marketed product batches can deliver a therapeutic effect consistent with that observed for product batches tested in the clinical studies.

Moreover, some advanced techniques can also be employed in the early stage of drug discovery to increase the success for identifying drug candidates. For example, high-content screening (HCS) is a cell-based screen method to investigate substances such as small molecules, peptides, and natural products that alter the phenotype of cells. Compared with conventional screening methods, HCS involves automated fluorescence microscopy with automated image analysis which can track the cellular morphology and intracellular parameters simultaneously at a subcellular level [45]. Therefore, HCS can provide a more comprehensive assessment of drug activity and safety in vitro, thus accelerating the drug discovery process. In addition, HCS can be coupled with 3D tissue culture, organoid, organ-on-a-chip, and other tissue engineering techniques to bridge in vitro and in vivo studies, providing a new window into drug discovery and development. Till now, although a number of assays have been performed to evaluate the activities of Ganoderma, none of these experiments are conducted under the HCS context. In future studies, for example, the antitumor activities of compounds, components, and fractions isolated from Ganoderma can be tested on the HCS and/or tissue engineering platform for more accurate screening and potential personalized precision medicine.

7.4.2 Development of Ganoderma-Related Health Products

In China, *G. lucidum* and *G. sinensis* are currently regulated either as herbal medicines or drug/health food materials, but not food materials. Thus none of the Ganoderma-containing foods can be produced by a common food manufacturer (with a QS production permit) and sold as a QS food on the market. They must get registration approval before being manufactured at health food GMP facilities and carrying a blue hat health food label when legally circulated on the market. Fortunately, China government just preannounced as of April 4, 2019, that it intends to regulate Ganoderma spores as a new food material with daily intake recommended 1–4 grams for adults. This rule, once formally approved by the national government, will allow products containing proper amount of Ganoderma spores to be made and sold on the market without preregistration approval (but a manufacturer still needs to submit quality standard documents to a relevant government agency). Ganoderma spores can therefore be made into coffee, candy, cookie, stick, cracker, capsules, tablet, or any convenient forms of food. With this progress, it can be reasonably expected that Ganoderma fruiting body will be treated as a new food material in the near future.

With the increasing consciousness of healthcare and growth of global nutraceutical industry, Ganoderma-related products have received great attention owing to ample scientific evidences for its health benefits, especially in the field of chronic and complex diseases. However, this industry confronts with significant challenges such as variety degeneration, low-quality control metrics, homogeneous competition of the products, and ambiguous understanding of active constituents and mechanisms of actions. It can be seen that there is still a big gap between traditional use and pharmaceutical application. Therefore, more scientific research should be taken for the modernization of Ganoderma industry by using more scientific thinking and state-of-the-art technologies. For the healthy, prosperous growth of Ganoderma industry, it is clear that the harmonious integration of research, development, production, and marketing by joint forces is important.

References

1. Lin ZB (2015) Chapter 1: historical data of Chinese Ganoderma (Lingzhi) research. In: Lin ZB (ed) Modern research on Ganoderma (Lingzhi), 4th edn. Peking University Medical Press, Beijing, pp 1–7
2. He BW, Xu DL, Ma L, Chen HJ (2016) Development and safety manufacturing of *Ganoderma* industry in Zhejiang. Edible Med Mushroom 24(6):353–357
3. Lin ZB (1979) Current situation of Ganoderma research in China. Acta Pharm Sin 14(3):183–192
4. Liu XM, Liu FP, Lin YH, Xu WJ (1991) Application of Ganoderma in drugs and supplements. Sub Trop Plant Sci 20(02):48–50
5. Lin ZB (2015) Chapter 4: cut-log cultivation of Ganoderma (Lingzhi). In: Lin ZB (ed) Modern research on Ganoderma (Lingzhi), 4th edn. Peking University Medical Press, Beijing, pp 56–73
6. Li MY (2015) Research and development of quality standard and specifications of *Ganoderma lucidum* from an industry chain perspective. Edible Med Mushroom 23(5):276–279
7. ISO 21315 (2018) Traditional Chinese medicine—Ganoderma lucidum fruiting body. International Organization for Standardization
8. Sanodiya BS, Thakur GS, Baghel RK, Prasad GBKS, Bisen PS (2009) Ganoderma lucidum: a potent pharmacological macrofungus. Curr Pharm Biotechnol 10(8):717–742
9. Paterson RRM (2006) Ganoderma–a therapeutic fungal biofactory. Phytochemistry 67(18):1985–2001
10. Bishop KS, Kao CH, Xu Y, Glucina MP, Paterson RRM, Ferguson LR (2015) From 2000 years of Ganoderma lucidum to recent developments in nutraceuticals. Phytochemistry 114:56–65
11. Hsu KD, Cheng KC (2018) From nutraceutical to clinical trial: frontiers in Ganoderma development. Appl Microbiol Biotechnol 102(21):9037–9051
12. Chen ZQ, Huang WL, Jin X, Liu ZM, Huang YJ, Li P, Zheng LY (2016) Research progress on Ganoderma lucidum intensive processing in China. J Food Saf Qual 7(2):639–644
13. Lawrence XY, Ilgaz A, Barbara A, Gregory A, Tara GB, Ashley B, Margaret C, David D, Joseph F, Adam CF, Scott F, Brian H, Henry H, Stephen WH, Robert I, Bruce DJ, Robert J, Paula K, Emanuela L, Sau LL, Richard L, Grace M, Mehul M, Ganapathy M, Moheb N, Roger N, Mary O, Thomas OC, Jim P, Raju GK, Mahesh R, Giuseppe R, Susan R, Anna S, Arzu S, Paul S, Vinod S, Ramesh S, Michael PT, Tony T, Bernhardt LT, Katherine T, Siva V, Martin VT, Fionnuala W, Russell W, Janet W, Geoffrey W, Larisa W, Louis Y, Diane Z (2015) Advancing product quality: a summary of the inaugural FDA/PQRI conference. AAPS J 18(2):528–543
14. Gou JH, Li MY (2013) Research of the operational mode of the agriculture-related whole industrial chain based on food safety: take Zhejiang Longevity Valley as an example. J Zhejiang Int Stud Univ 2(2):86–92
15. Kumar MNR (2000) A review of chitin and chitosan applications. React Funct Polym 46(1):1–27
16. Hayes M, Carney B, Slater J, Brück W (2008) Mining marine shellfish wastes for bioactive molecules: chitin and chitosan–part B: applications. Biotechnol J 3(7):878–889
17. Chen CC, Cheh LW, Yang JC, Tsai CM, Keh ES, Sheu MT, SU CH, Hong HH, Lin CT, Lee SY (2007) Non-shellfish chitosan from the fruiting body residue of *Ganoderma tsugae* for long-lasting antibacterial guided-tissue regeneration barriers. J Dent Sci 2(1):19–29
18. Su CH, Sun CS, Juan SW, Hu CH, Ke WT, Sheu MT (1997) Fungal mycelia as the source of chitin and polysaccharides and their applications as skin substitutes [J]. Biomaterials 18(17):1169–1174
19. Cheng YY, Qu HB, Zhang BL (2016) Chinese medicine industry 4.0: advancing digital pharmaceutical manufacture toward intelligent pharmaceutical manufacture. Chin J Chin Mater Med 41(1):1–5

20. Ferreira IC, Heleno SA, Reis FS, Stojkovic D, Queiroz MJ, Vasconcelos MH, Sokovic M (2015) Chemical features of Ganoderma polysaccharides with antioxidant, antitumor and antimicrobial activities. Phytochemistry 114:38–55
21. Cör D, Knez Ž, Knez Hrnčič M (2018) Antitumour, antimicrobial, antioxidant and antiacetylcholinesterase effect of Ganoderma lucidum terpenoids and polysaccharides: a review. Molecules 23(3):649
22. Ahmad MF (2018) Ganoderma lucidum: persuasive biologically active constituents and their health endorsement. Biomed Pharmacother 107:507–519
23. Nie S, Zhang H, Li W, Xie M (2013) Current development of polysaccharides from Ganoderma: isolation, structure and bioactivities. Bioact Carbohydr Diet Fibre 1(1):10–17
24. Zhang M, Cui SW, Cheung PCK, Wang Q (2007) Antitumor polysaccharides from mushrooms: a review on their isolation process, structural characteristics and antitumor activity. Trends Food Sci Technol 18(1):4–19
25. Li XL, Zhou AG, Li XM (2007) Inhibition of *Lycium barbarum* polysaccharides and *Ganoderma lucidum* polysaccharides against oxidative injury induced by radiation in rat liver mitochondria. Carbohydr Polym 69(1):172–178
26. Jiang H, Sun P, He J, Shao P (2012) Rapid purification of polysaccharides using novel radial flow ion-exchange by response surface methodology from Ganoderma lucidum. Food Bioprod Process 90(1):1–8
27. Xia Q, Zhang H, Sun X, Zhao H, Wu L, Zhu D, Yang G, Shao Y, Zhang X, Mao X, Zhang L, She G (2014) A comprehensive review of the structure elucidation and biological activity of triterpenoids from Ganoderma spp. Molecules 19(11):17478–17535
28. Li Y, Zhu ZM, Yao WX, Chen RY (2012) Study progress on triterpenoids from *Ganoderma Lucidum*. Chin J Chin Mater Med 37(2):165–171
29. Li L, Guo HJ, Zhu LY, Zheng L, Liu X (2016) A supercritical-CO2 extract of Ganoderma lucidum spores inhibits cholangiocarcinoma cell migration by reversing the epithelial–mesenchymal transition. Phytomedicine 23(5):491–497
30. Cheng YY, Qian ZZ, Zhang BL (2017) A strategy of constructing the technological system for quality control of Chinese medicine based on process control and management. Chin J Chin Mater Med 42(1):1–5
31. Newman DJ, Cragg GM (2016) Natural products as sources of new drugs from 1981 to 2014. J Nat Prod 79(3):629–661
32. Cao Y, Xu X, Liu S, Huang L, Gu J (2018) A cancer immunotherapy review. Front Pharmacol 9:1217
33. Lin TY, Hsu HY (2016) Ling Zhi-8 reduces lung cancer mobility and metastasis through disruption of focal adhesion and induction of MDM2-mediated Slug degradation. Cancer Lett 375(2):340–348
34. Lin TY, Hsu HY, Sun WH, Wu TH, Tsao SM (2017) Induction of Cbl-dependent epidermal growth factor receptor degradation in Ling Zhi-8 suppressed lung cancer. Int J Cancer 140:2596–2607
35. Wang K, Bao L, Ma K, Zhang J, Chen B, Han J, Ren JW, Luo HJ, Liu HW (2016) A novel class of α-glucosidase and HMG-CoA reductase inhibitors from Ganoderma leucocontextum and the anti-diabetic properties of ganomycin i in kk-ay mice. Eur J Med Chem 127:1035–1046
36. Wang K, Bao L, Zhou N, Zhang J, Liao M, Zheng Z, Wang Y, Liu C, Wang J, Wang L, Wang W, Liu S, Liu H (2018) Structural modification of natural product ganomycin I leading to discovery of a α-glucosidase and HMG-CoA reductase dual inhibitor improving obesity and metabolic dysfunction in vivo. J Med Chem 61(8):3609–3625
37. Su L, Liu L, Jia Y, Lei L, Liu J, Zhu S, Zhou H, Chen R, Lu HAJ, Yang B (2017) Ganoderma triterpenes retard renal cyst development by downregulating Ras/MAPK signaling and promoting cell differentiation. Kidney Int 92(6):1404–1418

38. Li X, Zhang JA, Huang J, Ma AQ, Yang JF, Li WM, Wu Z, Yao C, Zhang Y, Yao W, Zhang B, Gao R, Efficacy and Safety of Qili Qiangxin Capsules for Chronic Heart Failure Study Group (2013) A multicenter, randomized, double-blind, parallel-group, placebo-controlled study of the effects of qili qiangxin capsules in patients with chronic heart failure. J Am Coll Cardiol 62(12):1065–1072
39. Wang S, Hu Y, Tan W, Wu X, Chen R, Cao J, Chen M, Wang Y (2012) Compatibility art of traditional Chinese medicine: from the perspective of herb pairs. J Ethnopharmacol 143(2):412–423
40. Wang Y, Fan X, Qu H, Gao X, Cheng Y (2012) Strategies and techniques for multi-component drug design from medicinal herbs and traditional Chinese medicine. Curr Top Med Chem 12(12):1356–1362
41. Thomas DW, Burns J, Audette J, Carroll A, Dow-Hygelund C, Hay M (2016) Clinical development success rates 2006–2015. Biomedtracker/BIO/Bend/Amplion, San Diego/ Washington, DC
42. Lee SL, Dou JH, Agarwal R, Temple R, Beitz J, Wu C, Mulberg A, Yu LX, Woodcock J (2015) Evolution of traditional medicines to botanical drugs. Science 347(6219):S32–S34
43. Wang RL (2018) Analysis of registration for new Chinese herbal drug from 2011-2017. Mod Chin Med 20(7):910–914
44. Schmuff NR, Lin DT (2008) Chemistry, manufacturing and controls (CMC). In: Wiley encyclopedia of clinical trials. Wiley, Manhattan
45. Zanella F, Lorens JB, Link W (2010) High content screening: seeing is believing. Trends Biotechnol 28(5):237–245

The manufacturer's authorised representative in the EU is Springer Nature Customer Service Centre GmbH, Europaplatz 3, 69115 Heidelberg, Germany. If you have any concerns regarding our products, please contact ProductSafety@springernature.com

Printed and bound by CPI Group (UK) Ltd, Croydon, CR0 4YY
28/07/2026
02179159-0018